Henry Ford started America's largest and most prosperous industry: modern automobile manufacturing. He set up the system, and then he kept revolutionizing it.

A home-grown American genius, Henry Ford had all kinds of other unusual ideas that he had the power to implement. Harry Bennett was his man of all work, including the behind-the-scenes rough maneuvers that gave a leading gangster a Ford agency and canceled many of the benefits to Ford of his creation of the $5 day.

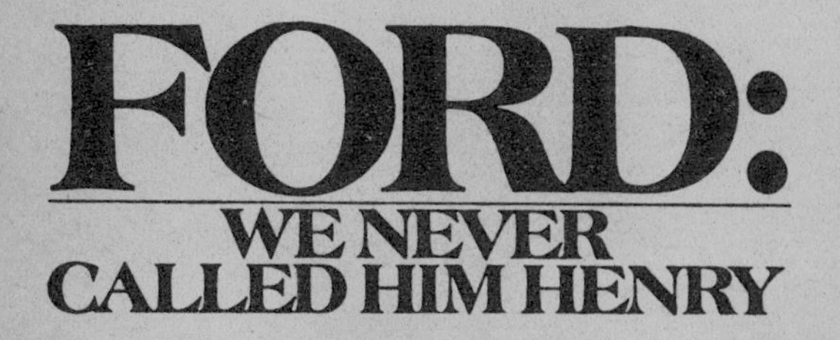

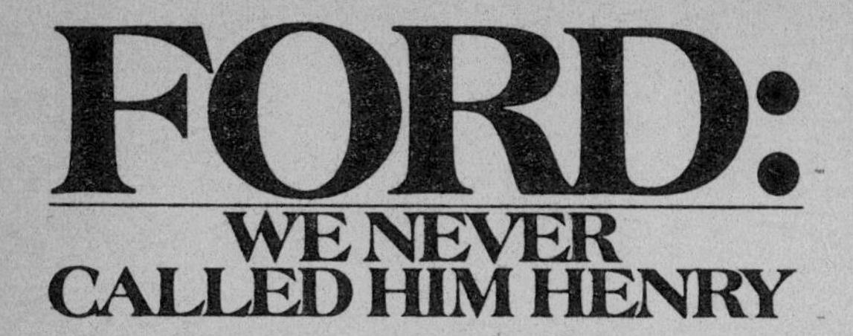

HARRY BENNETT

As told to PAUL MARCUS

A TOM DOHERTY ASSOCIATES BOOK

FORD: WE NEVER CALLED HIM HENRY

First Tor printing: January 1987

A TOR Book

Published by Tom Doherty Associates, Inc.
49 West 24 Street
New York, N.Y. 10010

Cover photo by Culver Pictures, Inc.

ISBN: 0-812-59402-9
CAN. ED.: 0-812-59403-7

Printed in the United States of America

0 9 8 7 6 5 4 3 2 1

Foreword

I WAS in Ann Arbor, Michigan, on an assignment for a national magazine, having dinner with a charming couple who were my "contacts" on the job.

"Of course," my host said, in explanation of a point he was making, "you know I used to be one of Harry Bennett's assistants at the Ford Motor Company."

I hadn't known; and the name Harry Bennett at once struck a spark of interest in me. I knew he had risen to a unique place in American industry, and that around him had grown up legends that were as profuse and fabulous as those that grew up about his employer.

"I'd like to meet that man," I said. "He must have a great story—if he wants to tell it."

"I'll ask him to see you," my host promised, "but I don't think he will. All his friends have urged him to tell his story, but he won't do it."

As it turned out, I didn't get to meet Harry Bennett. I finished my work and went back home. Several months passed. Then on a Sunday morning I got a phone call from my Ann Arbor hostess. She said, "We had dinner with Harry Bennett last night, and he's decided he wants to write a book. He's seen some of your work, and he wants you to come out and talk to him."

I spent six weeks with Harry Bennett. I listened to him talk and took exhaustive notes. I prodded and nagged at his memory, and asked countless questions. We ended, I felt, with material rich in insight into one of the century's great industrial figures.

As far as it is humanly possible, I have labored to keep out of this story any points of view of my own. I have only added some paragraphs of historical background, and these are set in italics so they may be recognized.

Since this book eventually must be evaluated by the historian of our era, I am willing to leave that task to him. He will, I trust, welcome the book to the growing shelf of Fordiana. No one was so close to Henry Ford as was Harry Bennett, and his story fills an important gap in our knowledge of our own times.

P.M.

Chapter 1

DURING the thirty years I worked for Henry Ford I became his most intimate companion, closer to him even than his only son.

This relationship between Mr. Ford and myself has been distorted in the public mind. It has been made to appear that Mr. Ford was a simple man who was merely ill advised—and that I was his advisor.

It wasn't true. After all, Mr. Ford was one of the world's greatest industrialists, a man who created a billion-dollar empire; no "simple" man could have achieved the things he did. The truth is, though I sometimes took issue with Mr. Ford, I was completely loyal to him. The reason I stayed with Mr. Ford for thirty years was that I always did what he wanted me to do.

It has been said, whenever anything happened at the Ford Motor Company that displeased the public, that I had done this thing, and without Mr. Ford's knowledge.

That isn't true, either. Nothing ever happened at the Ford Motor Company without Mr. Ford's knowledge; that was a physical impossibility.

I have been called a thug, a gangster, a pro-Nazi, an anti-Semite; it has been said that I was "fired."

All of these accusations are just plain lies.

I have no desire to eulogize myself. But for the sake of my family, and for my own peace of mind, I want to try to set the record straight.

I don't know any better way to do that than to tell the whole story from the beginning—from that day in 1916 when I first met Henry Ford.

Our first enlistment in the Navy had run out for both myself and a pal of mine, and we entered the port of New York, just in from Vera Cruz on the S.S. *Nashville*. This friend and I had joined the Navy together, had remained buddies throughout our service, and now intended to re-enlist. However, we meant to enjoy ourselves as civilians for a while before doing so.

Somehow we got involved in an argument with officials at the customhouse at Battery Park. I don't even recall the cause of the dispute, it was so trivial. But both of us were young and impulsive—I was twenty-four—and we got pretty hot-tempered. Soon, what had started out as an argument with both customs and city officials developed into a brawl.

My pal was a big man, well over six feet, who couldn't fight worth a darn, but was always ready to try. He laid out one official before he was himself subdued. I was small, being five feet seven and weighing only 145 pounds; but I wore a seventeen collar, and I had trained and boxed with some of the best in the Navy. I gave a good account of myself—until a big cop got me by the back of the collar.

By sheer coincidence, Arthur Brisbane, the noted journalist, was at the customhouse. A witness to the brawl, he scented a story and walked over to where the cop and I were threshing around.

"What's going on?" Brisbane asked the policeman.

"Oh, I've got a tough one here," the cop said.

Brisbane talked to my friend and me and got our story. By then we realized we were in trouble, and our only desire was to reach an enlistment office as quickly as possible and get back in the Navy, where, we thought, we would be safe from prosecution. Brisbane decided to help us. He took us to authorities who had the power to straighten the whole thing out. His prestige proved sufficient to get us cleared and out of our jam.

My friend decided on the spot that he'd had enough of civilian life, and promptly went back into the Navy.

To me, however, Brisbane said, "I'm going over to Broadway to meet Henry Ford, who is in New York, and I'd like to take you along, as you are from his home town."

Later, I wondered if perhaps Mr. Ford had told Brisbane he was looking for a young man who could take care of himself—or whether, having seen me in a scrap, he just decided on his own that Mr. Ford might like me. Whatever the truth, at the time I didn't reflect on the matter at all. I hadn't any objection to meeting Mr. Ford, so I went along with Brisbane just to see what it was all about.

When Harry Bennett was hired by Henry Ford in 1916, the River Rouge plant was being constructed. The Model T was already a huge success. Indeed, Henry Ford, in 1916, was perhaps better known to most Americans than their President. He had made one of the greatest contributions to modern times; he had conceived producing a standardized car for the common man at a time when the automobile was a rich man's toy. His famous "Model T," a rugged, mechanically simple, ugly black automobile, had put Americans on wheels by the millions. Seemingly impervious to abuse, built so that it could be repaired with a pair of pliers and a little baling wire, the Model T hauled the farmer's eggs and milk to market, and his family to the movies or to a distant relative's for Sunday dinner. It rolled along urban pavements in swarms, it ground up and down the mountains of both coasts, and slithered through the mire of the midwest plains, revolutionizing the social and economic life of a continent.

In 1905 Ford had started by building a racer that beat the competition. After the crash of a French racer, he realized that some of the parts of the European car, although lighter in weight, were stronger than his designs. Sorensen, who became Ford's amazing production chief, saw the new steel alloys as the fulcrum of the design for the Model T Ford. Henry Ford was a "hands on" participant. For example, Henry's suggestion that they slice off the top of the block made the single casting of the power unit a practical method of manufacture. Henry Ford made the major business decision to build his cars for "the great multitude."

When Brisbane and I arrived at 1710 Broadway, Ford sales headquarters in New

York, he at once introduced me to Mr. Ford. Mr. Ford was already in his fifties; he had sharp, gray eyes and heavy eyebrows, and was of medium height and spare build. He was alert and almost nervously quick. As I was to learn later, he was a grasshopper in his capacity for locomotion, and got in and out of a car with a jackknife motion that made men years younger seem awkward.

I wasn't much impressed. It didn't seem extraordinary to me that I had got into a fight, been rescued by a noted columnist, and met one of the world's most famous men all in one day. I don't know why, except that I was young and full of assurance, and not given to being impressed.

Brisbane told Mr. Ford about the brawl I'd been in. Mr. Ford chatted with me a while, asking questions about my background and my experiences in the Navy. Finally he said, "I can use a young man like you out at the Rouge."

I told Mr. Ford that I didn't want a job, that I intended going back in the Navy.

But he persisted in his offer.

Mr. Ford asked me, "Can you shoot?"

I said, "Sure I can," and with quickening interest wondered if he wanted me to be his bodyguard.

Then he said, "The men who are building the Rouge are a pretty tough lot, and I haven't got any policemen out there." And

I wondered if he wanted me to be a policeman.

As things turned out, both of these elements came into my work for Mr. Ford; but at the time, I didn't know what he had in mind—I only sensed excitement and adventure.

Actually, I was in conflict within myself about returning to the Navy. I loved Navy life. Yet when I was at sea I suffered severely from what was then called catarrh, but is now known as sinus trouble. So I decided to find out whether Mr. Ford had something for me that would keep me interested.

I said, "Well, I won't work for the company, but I'll work for you."

That seemed to satisfy Mr. Ford, and he turned me over to Gaston Plantiff, then New York sales manager. Plantiff was supposed to keep me busy in New York until Mr. Ford had time to get me set.

Mr. Ford left. I guess as far as Plantiff was concerned, I was just a pain in the neck that the boss had passed off on him. Either he didn't know what to do with me, or he didn't care much about the whole thing. Because as soon as Mr. Ford had gone, Plantiff gave me my instructions: He pointed to a big bunch of cars out on the sales floor and said, "Hang around there and don't let anyone see you."

The Ford Motor Company was successful

from the start. Early in 1908 Henry Ford decided to limit production to a single inexpensive design. Ford's company had discovered that the lower the price of their car, the greater their earnings. Ford began his amazing progress in gaining efficiency in production when he required that all the parts of his car be uniform so that they could be interchangeable between vehicles.

The Model T was strong, and a terrific performer, if somewhat plain. Its straightforward simple mechanical design could be fixed by anyone. Henry Ford conceived the overall design. He found brilliant designers like C. H. Wills who perfected the planetary transmission made from a special steel alloy which made driving a Model T different from driving any other make of car. The Model T was high-riding to accommodate to the terrible roads of that time. It gave a bone-shaking ride but was an efficient, sturdy, cheap car—ideal for small towns and farms.

Henry Ford's fabulous early success ran into the Selden patent claims that were being honored by the rest of the industry. Typically, Henry Ford fought in the courts—invalidating the patent. Ford was independent. He was not a joiner. His fight against the Selden patent claims was wonderful advertising for the Ford Motor Co.

In the early days cars were assembled by skilled mechanics. They started with the frame, adding each part until they had built a car—literally by hand. With Model Ts selling in record numbers, Ford found a gifted factory production expert, W. E. Flanders, who completely reorganized and streamlined the operations. All production records were broken. The existing plant at River Rouge became much too small.

The plant at Highland Park—both offices and factory—was built for Ford. Ford, along with his engineers, worked out new methods of assembly. He strove for straight-line production. The work was passed from one machine to another with few interruptions. But any hand method of passing articles along the line wasted time. Ford found that gravity slides speeded up the processing of small parts.

Ford's next solution was to set up numbers of the Model T chassis on wooden horses so that mechanics could move from assembly to assembly—making specialization feasible. These were axle, wiring, and motor specialists. Even such improved manufacturing methods could not keep up with the demand for the Ford cars.

In the subassembly plant for motors, axles, and magnetos, the work was moved by continuous belts that were fabulously effective in swamping the main assembly area with parts and subassemblies.

I hung around behind the cars. Plantiff didn't give me anything to do, nor did he put me on the payroll. I began to feel irked by the whole thing. I wanted to talk it over with Mr. Plantiff, but it was pretty hard to get any satisfaction out of him, as he was seldom in a condition suitable for conversation.

After a week or so, I thought the devil with this, and went to Detroit to stay with an aunt.

Mr. Ford had touched off my imagination, and I still hoped that working for him might offer enough adventure and excitement so that I wouldn't want to go back to sea. Somehow, I couldn't drop what had started out so promisingly.

Consequently, after a short time, I went out to Highland Park, then the main Ford plant, with a minor company official named Steinmetz whom I knew. Steinmetz got word to Ernest Liebold, Mr. Ford's secretary, that if Mr. Ford wanted to see me, I was around.

I had a talk with Liebold. Ernest Liebold at that time was very close to Mr. Ford and wielded great power in the organization. He was squat and heavy-set and had a short bull neck and close-cropped hair; for some reason, his coat collar always stood out about three inches from the back of his neck. Liebold and I were later to become antagonistic, but at the time, of course, I had no inkling that this was to be. Liebold curtly told me that Mr. Ford was too busy to see me, pointed out a desk, and told me to sit behind it.

Again a week went by, and I was given nothing to do. I was temperamentally incapable of sitting behind a desk and doing nothing. I had studied art, having got a taste for it from my talented mother, and liked to draw; so, to pass the time more agreeably while waiting for Mr. Ford to see me, I applied for a job in the art department.

While waiting to see the art director, Irving Bacon, I copied some illustrations out of a book I had at home called *Deeds of Valor*, which was a favorite of mine. When I got to see Bacon, I showed him the drawings.

"These are pretty good," Bacon said, scanning them, "but they look familiar, somehow."

I said, "Oh, I copied them out of a book called *Deeds of Valor.*"

"So that's it," he said. "I illustrated that book."

As soon as I got home again I looked at the book, and saw that he had, indeed.

Bacon put me to work in the art department. I got along all right, but I wasn't very happy. I was restless and finding it hard to get back into civilian life. To make matters worse, I incurred the hostility of Irving Bacon's superior, a man named A. B. Jewett. I guess Jewett didn't like the idea of my being around there at Mr. Ford's personal request. So he got a big fellow working in the department, who was about twice my size, to harass me in all kinds of ways.

I got disgusted pretty soon, and just quit and went home. I began angling for a commission in the Navy. But Mr. Ford sent for me. When I went to see him, he chatted with me a few minutes and then said, "Well, I just want to keep in touch with you," and that was all.

I went back to work again, but it was the same old story. There followed six or seven troubled, indecisive months. I guess at least five or six times more I quit and went home, and each time Mr. Ford sent for me, and each time he'd say, "Well, I just want to keep in touch with you," and that would be all. He made other appointments to see me, and then broke them. I didn't know what to do.

It's interesting to speculate now on what was going on in Mr. Ford's mind. Was he testing me in some way? Was he really too busy to get me set? Was he undecided about the whole thing? I don't know, to this day.

Anyway, things finally reached a boiling point. The hazing I was getting from Jewett's man got too much for me. I let him have it.

This wasn't the outcome Jewett had

expected, of course, and he wanted to fire me. I told Bacon, "I'm just waiting to get straightened out with Mr. Ford. Suppose I work in the darkroom, where Jewett won't see me?"

Bacon suggested this to Jewett, and he said, "All right, I just don't want any more slugging around the place."

So for a short while I worked in the darkroom.

Then Mr. Ford sent for me again. And this time it was the real beginning of my career with the Ford Motor Company.

By January 1914, two years before Harry Bennett was hired, Ford technicians installed an endless chain conveyor that took the whole assembly off the wooden horses and put it into continuous motion. The main conveyor was fed by endless overhead belts delivering parts and subassemblies. Henry Ford was the prime mover in this revolution in manufacturing methods which doubled production, while reducing the number of workers. Although self-taught and self-made, he was excellent at judging men and technical aptitude. His very success attracted the cream of managers and designers and he was generous in rewarding new ideas.

Chapter 2

MR. FORD told me he was going to send me over to the Rouge. He said, "I'm sending you over there to be my eyes and ears." Then he warned, "There may be a lot of people over there who want to fire you, but don't pay any attention to them. I'm the only one who can fire you. Remember, you're working for me."

We talked a while longer, and then I told Mr. Ford my troubles with Jewett's oversized stooge.

Mr. Ford said, "Well, we'll just lay a trap for that fellow." Then he explained, "I'll tell Bill Knudsen to ask the other executives during lunch for a big, tough guy to be sent over to the Rouge. I bet Jewett will send you."

I went back to the darkroom. A couple of days dragged by, and I thought they'd forgotten about me, and maybe I ought to go back to the Navy after all. Then I got a call from Knudsen. William S. Knudsen started with Mr. Ford in 1907; a big, jovial, easygoing man, he had built fourteen of the company's branch factories and was now in charge of building the Rouge plant. He said, "Mr. Ford wants you out at the Rouge. Do you know your way out?"

I told him I didn't, and Knudsen said he'd have Jewett send me out.

Much later, I learned that Mr. Ford's plot had worked out perfectly.

Every day Ford executives met for lunch in a pine-paneled corner room at the Dearborn Engineering Laboratory, usually with Mr. Ford. These lunches, where executives gathered at a big round table, were really meetings, and lasted anywhere from one to five hours. At this lunch table, then, Knudsen, acting on Mr. Ford's instruction, said, "I need a big, tough guy out at the Rouge. Do any of you know one you could send out?"

True to Mr. Ford's prediction, Jewett fell all over himself volunteering to send me.

It should be clear that Mr. Ford had in no way intimated to Knudsen that he had me in mind; and when Jewett volunteered my services, Knudsen just assumed that I was, indeed, a "big tough guy."

Irving Bacon drove me out to the Rouge and left me at the gate.

Highland Park was the core of the Ford Motor Company in 1917; it was there that Ford was reaching his goal of producing a car every minute. The Rouge then consisted of only a blast furnace and "B" building. But the great Rouge assembly line had already been designed by Ford's architect, Albert Kahn, and was then under construction. Thousands of construction workers swarmed over the vast acreage of the Rouge, a hard-bitten crew. The automobile industry was no place for hothouse plants and the Rouge in particular was tough, brawling, violent.

I had no idea where to go, and turned to a giant Polish foreman I saw. I said, "Where can I find Mr. Knudsen?"

"Why do you want to see him?" the man said.

"That's my business," I announced.

Without another word the big Pole clipped me on the jaw and knocked me down.

I sat there, shaking my head to clear it. I thought: Mr. Ford has framed me. I did not know I was the butt of a Knudsen witticism. Believing me to be the "big guy" he had asked for, Knudsen had advised this man, "If he gets fresh, clip him." He thought, of course, that the fellow would never dare swing on me.

The big fellow stooped over and put his hands beneath my arms. "The next time you're asked a question around here," he said, "don't get so cocky," and he helped me up.

"Thanks," I said. Then I pushed his chin up with my left and swung with my right. Only his jaw didn't come down, as I had expected, and I hit him in the neck. The blow not only laid him out, but left him speechless as well. He went around the Rouge whispering for weeks.

I proceeded to Bill Knudsen's office, where I announced myself to his clerk. I cooled my heels there for some time. As I learned later, the story of my fight preceded me into Knudsen's office. When the clerk at last told Knudsen I was waiting to see him, Knudsen said, "Oh, Bennett—a big fellow?"

"No, I wouldn't say so," the clerk said.

Puzzled, Knudsen came out to the waiting room himself. By now it was late, and I was sitting there alone. Knudsen looked at me. He clapped his forehead with his hand and exclaimed, "Jesus Christ!"

There wasn't much in my early life to prepare me for my career with the Ford Motor Company.

I was born in Ann Arbor, Michigan, on January 17, 1892. I never knew my father. While I was still an infant he was killed in a brawl with a man who struck him over the head with a chair. I don't know the reason for the fight; some said it was over a woman, others said my father took objection to the way this man was treating my grandmother.

My mother was a fine, cultured woman, a talented painter, liked and admired by all who knew her. She was a great influence in my life. Many years after my father's death, my mother married again. Her second husband was a professor, Robert Winslow, who taught in the engineering school of the University of Michigan in Ann Arbor. My stepfather and I never got along too well.

There were good times and hard times in my childhood. I had my boyhood fights and my boyhood fun. I sang in the St. Andrew's Episcopal Church choir. After a falling out with the pastor, my mother and I changed to the Church of All Creeds in Ann Arbor, attended by Protestants, Catholics, and Jews—an experience that molded my attitude toward faiths other than my own. My stepfather died a few years after he married my mother, and for a while I was "orphaned out" to one of my uncles, who lived in Saginaw.

When I was fifteen years old my mother and I moved to Detroit. My ambition then was to be a commercial artist. My mother entered me in the Detroit Fine Arts Academy, thinking it would do me good.

Shortly I came into conflict with my mother. She was advised by an uncle of mine, Sid Bangs, a well-known songwriter then, that I was "spoiled." My mother began tightening up on everything. This led to surliness on my part. I felt she was holding out on me and got angry about it. When I was seventeen I ran off to join the Navy, with the friend I mentioned earlier.

My time in the Navy was nothing unusual. My experience represented that of the average fellow in the Navy who is happy enough where he is. I had no ambitions as an enlisted man. I did, however, have ideas of getting into Annapolis. I had got the notion from ensigns I'd met that you didn't have to know too much to get out of Annapolis.

I did a little of everything in the Navy. I drew cartoons for a Navy magazine. Toward the end I did some routine diving, but no more than hundreds of others; it was never a career with me. I worked for quite some time firing with the Black Gang on the *Texas*, the *Ohio*, and *Nashville*, and passed coal on the *Ohio*. I learned what hard work meant. I'd always wished that I were a big man and all my life I took on jobs that were meant for a man bigger than I.

I became interested in boxing. Because I

was small, it was the only sport for which I found myself eligible. I came to love the sport, and seemed to be well equipped for it. I never had any physical fear of anyone, and I have always been ready to fight if I thought I was being pushed around.

I trained and boxed with some of the best in the Navy, as I have already said, but the long circulated tale that I was a champion in the Navy is ridiculous; there were too many fine champions in the Navy for me to claim to be in a class with them. I never fought for money in my life, and never considered I was good enough to become a professional.

In later years, I helped many boxers. They, and others who liked me, with only the best of intentions built me a completely erroneous reputation as a champion fighter. I tried to knock down this phony story, but all I got for my efforts was the response that I was being modest.

Actually, my career in the Navy was a combination of a vacation and a sightseeing tour. There was never a dull day in the Navy, and that's why I loved it. Probably there were few jobs that could have kept me in civilian life for long, other than working for Mr. Ford.

Henry Ford loved machinery. He was never in his office. He circulated through the plant observing with an expert eye—encouraging with the useful suggestion. He would stay in the plant all night if important changes were being made—and he usually came to work at 7 AM. At all times he knew what was happening in every department.

Exactly what was my job with Mr. Ford?

Well, I guess the simplest way to define my position is to say that I was Mr. Ford's aide, his man of all work.

Titles weren't popular at the Ford Motor Company. In fact, from 1918, when Mr. Ford gave the presidency to his son, Edsel, until 1943, when he assumed it again, he himself had no title at all, though he was a director. In fact, most of the years I was with Mr. Ford there were only two titles in the whole company—a president-treasurer and a secretary–assistant treasurer. It was a one-man show, and Mr. Ford was jealous of authority.

A magazine writer who visited the company in the thirties reported: "It is a little difficult to sort out the Ford executives for the reason that Mr. Ford does not believe in titles. . . . Ask any of the executives what their jobs may be and after a few minutes' thought they may admit that they look after this or take care of that. But never say they: 'I am the Sales Manager' or 'I am the Chief Engineer.' "

But in addition to this general attitude toward titles, *I* had no title because it would have been physically impossible for me to hold down any one job at the Ford Motor Company, because of the way Mr. Ford kept me with him all the time, and the way he had me jumping around from one task to another.

If I had worried earlier that I might not find enough excitement with Mr. Ford, I didn't have to worry long.

Here are some of the things that kept me interested.

Mr. Ford, always concerned for his personal safety, had a small-arms target range over his garage at the Residence, and almost from the beginning he took me up there and we had target practice together.

I had kept my diving suit from the Navy, and carried out several diving assignments for Bill Knudsen.

Once we were were tipped off that a gang meant to hold up our pay car on the D.T. & I. Railroad, then owned by the company. A fellow named Milt Johnson, recently Twin Cities branch manager, and I captured the bandits. We parked our car near the tracks in some pretty rugged country, where we knew the hold-up attempt was to be made. When the bandits showed up, we chased them. I drove the car and Johnson hung on the running board and blasted away with a shotgun. After a wild pursuit the bandits threw their guns out and surrendered.

Again, there was the time, somewhere

around 1920, when a Detroit News reporter named Art Ogle brought me a tip that there was to be a robbery attempt on our pay office at the Rouge. Ogle got the tip from a West Side mob in Detroit, who tipped him off either because they thought they'd be blamed or because they'd been left out—we didn't know which was the reason.

When the car we were expecting drove down Miller Road, a public thoroughfare that bisects the Rouge plant, I went out to meet them. The car was driven by a man named Walters.

I said to them, "There's a whole arsenal in there waiting for you. If you go in, there are going to be a lot of people killed, and some of them will be you."

I also told them there was a newspaper reporter and a police judge watching them—which was the truth—but that if they turned around and went away, nothing would happen to them.

I guess they weren't any more anxious to be shot at than anyone else would be, and were glad of a chance to get out of it this easily. But at first they suspected some kind of double cross. They wanted me to ride on their running board until they had cleared the plant. However, I convinced them that if they forced me to get on, the shooting would start.

Finally they believed me and drove away.

I went inside and said to Ogle, "Well, that's that."

"Oh," he said, "just like that?"

He was mad. He'd brought me the tip because he thought he was going to witness a gun battle and go back to his paper with an eyewitness account. "There's a great story," he said, "going right down the street."

It was things like these that kept me there, not missing the Navy at all.

Chapter 3

FORD *was not only a "hands on" leader. He was daring and ambitious, he had apparently inexhaustible energy, and he was a man of amazing vision. The River Rouge plant was the largest manufacturing facility in the world—starting from scratch to finished product. It had the primary means—blast furnaces and cooking ovens—to use iron ore and coal. But this was not enough for Ford. He wanted to own everything that went into his cars. He purchased iron and coal mines. He had his own fleet of ships for transporting raw materials. He even grew rubber on land he owned in Brazil.*

The first nine or ten years I was with Mr. Ford was an educational period for me. I learned to know Mr. Ford. I learned how he thought, how he felt, how he operated. I also learned what he expected of me.

In those early days, he once said to me, "Harry, never try to outguess me."

I was green then, and I didn't quite get it. I said, "You mean I should never try to understand you?"

"Well," Mr. Ford said, "that's close enough."

It bothered me. He had said it with an air that led me to believe he laid considerable weight on the advice, and yet I wasn't quite sure of his meaning. So I brought it up again a few days later. "Mr. Ford," I said, "did you mean you didn't want me to try to understand you?"

"Yes," he said again, "that's close enough."

What he meant, I eventually realized, was that he expected me to carry out his wishes without probing for his motives. Mr. Ford always had a motive for everything he did; usually he had two motives—the one he gave, and the real one. He didn't want me digging into that too far.

Nevertheless, I learned a great deal about him. Mr. Ford was a great man; but he had his own peculiarities.

First of all, I learned, Mr. Ford was inconsistent. He was the most inconsistent man I ever knew. Yet he hated nothing more than being called inconsistent. I used to tell Mr. Ford that he was inconsistent, and he'd get angry and say, "If you mean I change my mind, I reserve the right to change my mind any time I want to." Just because he said something yesterday never meant that he felt committed to it today.

Many times in critical situations his attitude was: "Harry, let's you and him have a fight."

He never wanted to have people hurt; or, if they were hurt as a result of his orders, he didn't want to know about it.

Mr. Ford always claimed that he didn't care what people said about him, but he certainly wanted to *know* what people said. One of the tasks he gave me, over and over for thirty years, was finding out what people thought of him. He was insatiable for information.

Yet Mr. Ford hated to have strangers get confidential with him. If someone he didn't know very well sidled up to him at a social affair and began feeding him some gossip, Mr. Ford got away as fast as he could. Later he would tell me, "So-and-so was piddling in my ear." That was a favorite expression of his.

Mr. Ford was extremely superstitious. If he put on a sock inside out in the morning, he'd never change it. He was afraid of black cats, walking under ladders, breaking mirrors, and all the rest of it. But he had ways of rationalizing these things. He'd say, "If a black cat crosses the road and you're superstitious, then you'll drive more carefully, and that's a good thing. Anyone who will walk under a ladder deserves to get a paint pot on his head." And on a Friday the thirteenth, you could hardly get him to move.

As far as loyalty was concerned, Mr. Ford didn't seem to care too much. I saw men disloyal to Mr. Ford over and over, and yet if they were useful to him, he'd keep them on. He never let emotion interfere in business.

The one thing in the world Mr. Ford couldn't stand was ridicule. He couldn't stand any kind of slur on his intelligence. While I was with him I sometimes ignored orders; I took issue with him any number of times; but I know now that if I'd ridiculed him, I'd have been through.

During those early years Mr. Ford gave me a lecture on the matter of gifts. He told me, "Never give anything without strings attached to it."

A gift from him always had a rubber band on it. Gifts to me were the one exception.

Time after time I saw Mr. Ford give things to people—cars, tractors, and what-not—and then if he became angry at the recipient of the gift, he'd take the gift back again, without any explanation. He'd never give a title or deed with a gift; he'd tell the person, "Now, this is yours for life." That was a favorite expression. But when he got angry, he forgot his promise.

The best example of this sort of thing I know is the car he gave to a man named Ash.

Ash ran the power plant at the Rouge. He was an old, loyal employee. In fact, he had worked with Mr. Ford at the Edison Electric company as a fellow mechanic before Mr. Ford ever began making cars. He was a kindly, somewhat simple man who had no thought but for his job.

Mr. Ford decided to reward Ash for his many years of service, and gave Ash a new Ford car.

Ash was terribly grateful. He wanted to show his appreciation. So whenever word traveled down the grapevine that Mr. Ford was coming to the powerhouse, Ash would grab a rag and run out to where his car was parked and start polishing it with loving care.

Mr. Ford noticed this, all right, but the performance didn't have the effect that Ash had expected. It made Mr. Ford angry. He came to me and said, "That fellow Ash hasn't done a lick of work since he got that car. You go and take it away from him."

I just ignored this order, as I did others of a similar nature. But a few days later Ash's car disappeared.

Ash was wild. Of course, he had no idea what had happened to his prized possession. He notified me of the theft, and he also notified the Dearborn police.

Ash never got his car back, to my knowledge.

It wasn't often, though, that Mr. Ford came right out and told you something he thought you should learn, as he did with me in the matter of gifts. He had his own unique educational method, and I ran up against it very early in my career with him.

The following incident, which happened because Mr. Ford wanted a slag separator installed at the Rouge, has been told before—but never accurately.

In 1920, Mr. Ford bought the Detroit, Toledo & Ironton Railroad, a dilapidated, vest-pocket line that he put into first-class operating order. One day when Mr. Ford and I were together he spotted some rust in the slag that ballasted the right of way of the D.T. & I. This slag had been dumped there from our own furnaces.

"You know," Mr. Ford said to me, "there's iron in that slag. You make the crane crews who put it out there sort it over, and take it back to the plant."

There were miles of this stuff, and to me Mr. Ford's order was a joke. I could plainly see, as anyone could, that the reclaimed iron would never repay the labor required to do the job. However, I figured Mr. Ford owned the plant, and if he wanted to do something silly like that, it was his own business. So I started having it done. I got a crane out there, and over a hundred men, and put them to work.

The men were bringing some of this stuff back to the slag crusher when Charley Sorensen saw them.

Charles E. Sorensen was then one of the three people who together headed Ford production—himself, Peter E. Martin, and Knudsen. Sorensen, sometimes known as "Cast Iron Charlie" because of his ability at casting, was a Danish immigrant who started with Mr. Ford in 1904. He began as a patternmaker and did some of the work on Mr. Ford's famous racer, "999," which Mr. Ford and later Barney Oldfield drove to fame. Sorensen was about six feet tall, heavy but well proportioned, always well groomed, and a handsome man; people used to say he could have been a Hollywood star. Sorensen went up to some of the men who were bringing in the slag, and demanded, "What the hell's wrong with your heads?"

They told him I'd given them orders to do it.

Sorensen, whose headquarters were then at Dearborn in the Highland Park plant, came to see me. He asked me if I didn't think it was a great waste of money to have this work done.

"Yes, sure it is," I said.

Sorensen exploded, "Then why in hell are you doing it?"

"Because Mr. Ford told me to," I said.

"I suppose," Sorensen said, "if Mr. Ford told you to tear down that powerhouse over there, you'd tear it down."

I said, "If this is a game, just be sure Mr. Ford doesn't tell me to tear it down, because it will come down, all right."

I thought that Mr. Ford had wanted to make me look foolish. I didn't lose any time getting to him with this, and told him what had happened with Sorensen.

But Mr. Ford only said, "Now, Charley isn't like that at all. He's just a pinch-penny on things like this." Then he explained, "But what you did will save us a million dollars. Now Charley will get a separator to take the iron out of the slag before it goes out there."

I wasn't cooled off any, and I said, "Well, why didn't you just tell him to do that? What is this, a three-ring circus?"

Mr. Ford saw that I felt I'd been made a fool of, and that I was angry enough to quit. The next morning he called for me to take me to work, and kept pretty close track of me for the next three days. I didn't get far out of his sight.

In the future, I taught Mr. Ford not to fool me. After that, when he gave me one of those phony orders, I'd get started so fast there'd be a lot of damage done before I could be stopped.

I must admit that Mr. Ford's scheme worked. A slag separator was installed at the plant. But that was his educational method. He'd spend thousands of dollars letting a man go out on a limb rather than give a simple, direct order.

As for Sorensen, I don't think this whole thing bothered him very much. He was similar to Mr. Ford in that he was a cold-blooded executive. Sorensen was one man who was never bluffed by Mr. Ford. He usually got what he went after, even when it was against the wishes of Edsel, Mr. Ford's only son.

During those early years, a great deal of my time was taken up by making investigations into marital situations for Mr. Ford.

One of his peculiar traits was that he was forever trying to reunite families that had broken up, without considering the desires of the couple involved. If there was a divorce or a separation, particularly among his relatives, then right away Mr. Ford wanted to fix everything up and bring the estranged couple back together again—and never mind what *they* wanted.

Needless to say, this kind of "help" was seldom welcome, and he didn't always realize what effect he was having on people. For instance, Mr. Ford's stubborn efforts to get a nephew to go back to his wife led him into actually persecuting the man.

In few instances did Mr. Ford's efforts ever work out. Once in a while couples did come together again but usually it was only briefly.

Having interfered in the private lives of most of his wife's relatives, Mr. Ford decided he wanted me to locate all the Littigots. They were his mother's side of the family—his father was an orphan, and no one knew anything about his family. Mr. Ford said to me, "I've done something for all the Bryants (his wife's family)—now I'm going to spend just as much on the Littigots."

When I did manage to dig them up, however, the Littigots turned out to be independent people. All they wanted of Mr. Ford was to be let alone. Mr. Ford, running true to type, found a broken marriage and at once tried to patch it up, although the man and woman detested each other and were quite happy to be divorced. The wife said, "Go back to that fat slob? Not for a million dollars." And the husband, who had married again, said, "I'm in love with the wife I've got!"

So in the end, with the Littigots as with others, all that our efforts accomplished, though well intentioned, was to develop a lot of resentment toward Mr. Ford.

But while Mr. Ford's relations with his relatives were strained, they were as nothing compared to the difficulty of his relations with his only son, Edsel.

Chapter 4

IN 1918 Mr. Ford resigned as president of the Ford Motor Company, though he kept his seat on the board, and Edsel, then twenty-five, was elected president.

For a couple of years Mr. Ford had been trying to buy out his minority stockholders. They were not of a mind to climb out of their comfortable seat in the Model T. But when Mr. Ford announced that he was leaving the company to devote his time to "other interests," this put his stockholders in such a panic that they sold out to him pretty much on his own terms. However, none of them had much cause for complaint; on the basis of small original investments, they left the Ford Motor Company as multimillionaires.

Though Edsel remained president of the Ford Motor Company from then until his death, Mr. Ford never really gave him anything but the title. Mr. Ford was boss, and no one ever had cause to doubt it.

Edsel, who had a mild voice and quiet manners, was built like his father—slender and long-legged. He was a nervous man; when he got angry, he threw up. He was just a scared boy as long as I knew him. Mr. Ford blamed himself for this. He had always overprotected Edsel. He had had Edsel privately educated, refused to let him go to college, and had taken him into the business and under his wing when Edsel was about eighteen. All this undoubtedly had its effect on Edsel's character.

I know that Mr. Ford loved Edsel dearly, but there was an unrelenting though sometimes hidden struggle between Mr. Ford and Edsel that went on to the day of Edsel's death. The base for this was in Mr. Ford's desire to have Edsel be just like him—to live like him and to think like him.

Whenever Mr. Ford thought someone was selling Edsel ideas contrary to his, he came down on the offender like a ton of bricks. If Edsel began carrying out an idea that Mr. Ford didn't like, Mr. Ford never said a word to him; he let Edsel go ahead, and meanwhile worked against the project himself. It was his idea of the best way to educate Edsel and get some sense into him.

At one time Bill Mayo, chief engineer of the Ford Motor Company, convinced Edsel that some new coke ovens were needed. Edsel told Mayo to go ahead and build them.

Mr. Ford disapproved, but he did nothing to stop it. He came to my office and told me, "Harry, as soon as Edsel gets those ovens built, I'm going to tear them down."

I saw nothing funny in this, and I promptly went to Edsel and told him what his father had said. But Edsel resented my telling him. He said, "Bill knows more about coke ovens than Father. I don't think he'll do anything of the kind."

There wasn't anything more I could do, except wait for the explosion. It came. As soon as the ovens were finished, Mr. Ford had them torn down.

This feeling of hostility on Mr. Ford's part toward people he thought were influencing Edsel carried over even to Edsel's relatives. In 1916 Edsel married Eleanor Clay and built a home on Grosse Pointe. Grosse Pointe is a suburb of Detroit that is largely inhabited by the "aristocracy" of Detroit. It is also the home of most of the General Motors people, many of whom were Edsel's in-laws and friends. Mr. Ford was so jealous of these people, and their real or imagined influence on Edsel, that he carried on a bitter feud with Grosse Pointers for thirty years. He made repeated, if fruitless, efforts to get Edsel to move to Dearborn, where his own residence was located.

One of the people against whom Mr. Ford vented his wrath was Ernest Kanzler, who was married to Eleanor Ford's sister. Kanzler was an executive in the company, and was a capable man. Mr. Ford himself told me that Kanzler had never asked for a job with the company—Mr. Ford had invited him in. But bad feeling developed between them because Mr. Ford got the idea that Kanzler was influencing Edsel away from Mr. Ford's ideas.

Mr. Ford told me to "bounce" Kanzler, and "make as much noise" as I could. But I felt that Mr. Ford was just blowing his top. George Brubaker, who was Mr. Ford's brother-in-law and office manager of the plant, and I talked Mr. Ford out of it.

However, Mr. Ford didn't give up his idea of getting Kanzler out. A short time later he called me up at home. He said that Kanzler had just fired Sorensen, and what was I going to do about it?

Mr. Ford knew my admiration for Sorensen. He thought that my own anger would be so great that I would do as he wanted—fire Kanzler, and "make a lot of noise" doing it.

Actually, Kanzler *had* fired Sorensen. But Sorensen took this "firing" as a joke, and went to Mr. Ford and laughed about it. It would have taken a lot more than Kanzler, then, to get Sorensen out of there.

I didn't know all this at the time of my conversation with Mr. Ford. I took his statement at face value. But I still didn't want to get involved in any Ford family disputes if I could help it. I told Mr. Ford that, since Kanzler was Edsel's relative, it was a tough one for me to handle.

"All right," Mr. Ford said, suddenly giving up any pretense that Sorensen had been fired. "I'll tell Charley to fire him."

I guess he thought that arrangement was at least poetic justice.

Presumably Sorensen didn't have many qualms about carrying out this assignment; anyway, Kanzler left the company.

Edsel's wife came to Mr. Ford after this and wept, and the whole family carried on quite a bit. However, Edsel, his wife, and other members of the family became reconciled, and the repercussions of the feud that had so long strained the ties of the Ford family died away, and there was peace in the family once again.

Nevertheless, I believe that Kanzler's

leaving marked the parting of the ways between Mr. Ford and Edsel. Things were never again the same between them. When Edsel wasn't deeply embarrassed by the things Mr. Ford did (which was often), his attitude was one of cynicism.

Mr. Ford always kept track of everything Edsel did. Countless times he asked me to check on Edsel and his family. I refused point-blank. Mr. Ford was put out about this, and used to go around telling people that I refused to check on Edsel for him, apparently expecting the world at large to share his indignation.

Nevertheless, Mr. Ford did constantly get information on Edsel, although for years I didn't know how. Then one day I caught Mr. Ford and one of Edsel's servants in a conference behind a building at the Dearborn laboratories. The servant ducked when he saw me. I asked Mr. Ford what they had been talking about, and he admitted that the man had been keeping him informed of Edsel's private affairs.

I was at Edsel's residence only once in my life. That was during prohibition, and under rather difficult circumstances.

At a time when Edsel was out of town, Mr. Ford asked me to drive him to Edsel's home. On the way out he told me that he'd heard Edsel had a stock of whisky and champagne, and he was going out there to break it all up.

Mr. Ford was violently opposed both to drinking and to smoking cigarettes. The prejudice against cigarettes he got from his old friend Thomas A. Edison, whom he admired greatly. Smoking was never permitted around the Ford plant.

When we arrived at Edsel's home, I refused to get out of the car. Mr. Ford threatened me, as he did often in those early days, saying, "Well, maybe you should have gone back to the Navy, after all." But I stayed in the car anyway, and he went in alone.

I don't know what Mr. Ford did while he was in the house. But when he came out and got into the car again, his clothes smelled of liquor.

Edsel and I were never on completely friendly terms. Mr. Ford periodically would make efforts to bring us together. Then, just when Edsel and I were making progress toward a friendlier relationship, Mr. Ford would break it up. He'd say to me, "Now, Harry, you think you're getting along all right with Edsel, but he's no friend of yours."

Despite all the difficulties we had between us, I always refused to interfere in Edsel's life, as I've said. On Edsel's part, he knew of serious, blundering mistakes I made, yet he never told his father about them.

Up to 1916 Henry Ford relied on his

secretary/treasurer James Couzens to carry out any abrasive aspects of management. Besides being a master salesman, Couzens ran the purchasing department. He dominated the Ford traffic department. Henry Ford was able to concentrate on production because Couzens handled the front office brilliantly.

Couzens contributed, with Ford's full support, the radical idea that the Ford Motor Company would increase its earnings by doubling the going wage scale of $2.50 for a nine-hour day. Ford paid $5 for an eight-hour day. Spending more on wages was, as Henry Ford said, "one of the finest cost-cutting moves we ever made."

The unexpected result was a riot when ten thousand men attempted to apply for jobs. Fire hoses were used in zero-degree weather. Ford had to think about plant and personal security. He picked Harry Bennett as his security officer.

But to get back to those early days again—a number of things happened that were important.

In 1919 Mr. Ford bought a small newspaper, the *Dearborn Independent*, and set up editorial offices for it in the tractor plant in Dearborn.

E. G. Pipp was the first editor of the *Independent*, and a man named William J. Cameron was his assistant. In its early issues, the paper came out strongly against sin. It was also strongly against "Wall Street" and "capitalists."

"Why, you're a capitalist," I told Mr. Ford.

"No, I'm not, either," he insisted.

The $5 day had an impressive economic impact. Between 1914 and 1916 homes of Ford employees rose in value from just over 3 million to 20 million dollars. Workers' savings increased from less than $200 to $750 per individual. Ford's methods had national impact, revolutionizing industry and bringing affluence for a whole society.

Also in 1919, Mr. Ford went through a lawsuit with the *Chicago Tribune* that won great notoriety in its day. I had no connection with this at all, and so I only know the gist of it.

The case came to trial in the summer of 1919 at Mt. Clemens, Michigan, a county seat outside Detroit. Mr. Ford was put on the witness stand, and in an effort to discredit him the *Tribune*'s lawyer ridiculed Mr. Ford bitterly and at great length. He managed to show that Mr. Ford was poorly educated, that he actually didn't read too well, and that his vocabulary was pretty limited. He didn't succeed in making a fool of Mr. Ford, by any means, but it was a bitter and humiliating experience for him, and it had at least one important consequence.

Mr. Ford told me, when it was all over, that he would never, under any circumstances, appear again on the witness stand.

Ford made a point of employing the handicapped. In 1919, out of a work force of almost 45,000, 9,500 had some kind of handicap. The handicapped got the full wage, with profit-sharing. At the same time Ford employed around 500 ex-convicts.

Between 1919 and 1921 there was a radical shift of executive personnel in the company. Most of those shifts would be of little interest here, but two are important.

In 1921 Frank L. Klingensmith, treasurer and a vice-president of the company, left. Klingensmith helped me a great deal during my first years with the company, and had always been friendly to me. His leaving was of significance because it gives an insight into Mr. Ford's relations with his executives.

It was well known in the industry that he never paid his executives very well. I don't think any of them received salaries comparable to those paid elsewhere in the industry. Their financial situations were made even more difficult by the fact that Mr. Ford hated his executives to have interests outside the company. If they did, and he found out about it, he called them crooks and got rid of them.

That is what happened to Klingensmith.

Klingensmith had a lot of outside interests—notably, stock in other companies. Liebold learned of this and informed Mr. Ford. As was the case with other executives, Mr. Ford didn't like it. He told Klingensmith of his information, and Klingensmith readily admitted holding stocks in other companies. Mr. Ford then told Klingensmith he'd have to chose between the Ford Motor Company and his outside interests. Klingensmith cheerfully chose his outside interests, and left the company. So far as I know, he and Mr. Ford parted on friendly terms.

Knudsen left the Ford Motor Company in 1921 and went over to General Motors, to head the Chevrolet division—as far as I know, only because he wanted to. With his leaving, Sorensen became top man in production.

Once Sorensen came into complete control of production, he was not only the driving force that got things done, he was the cement that held the organization together.

In all the years I knew Sorensen he was never a politician in the plant. He was a cold, aloof man, and never had any social relations with anyone in the company—myself included.

Sorensen never held a grudge. He might be fighting mad at a man one day, but if that fellow was getting along all right in his job the next day, Sorensen would forget it.

Part of Sorensen's great value to the company was that he could always visualize the finished product when something was under discussion. He was capable of giving quick, positive decisions. But above all, Sorensen really got things done in the plant. He was a driving, efficient executive who thought of nothing but the car.

Sometimes men had nervous breakdowns, trying to keep pace with him. When some fault in a car was in the process of being corrected, Sorensen went at the job at a brutal clip. If a man was made of tough stuff and stuck it out, he was all right. But if he broke down and had to go away for a rest, Sorensen simply dealt with the fellow who took over his work and was rugged enough to stick it out. When the first man came back, he found out he wasn't exactly fired, but he didn't have a job.

Very often the returned executive just quit. But if he asked for a different job or the one he'd had before his promotion, he'd get it.

There were stories circulated that when Sorensen was displeased with someone, he would notify that individual of his displeasure by taking an ax and chopping up the offender's desk. The stories were untrue.

But Sorensen did have a brusque manner, and never wanted thanks for what he did.

I remember one time Sorensen promoted a man, and did it in such a way that the fellow thought he was being fired. "You leather-headed old so-and-so," Sorensen said, "you're the only one around here who knows anything." It sounded like brutal sarcasm.

The man quit and went home. I called him back and explained, "Why, man, Sorensen was promoting you."

"I don't give a damn if he was," the man fumed, still boiling mad. "He can take his damn job and shove it." And he never did come back to work.

Mr. Ford depended heavily on Sorensen. He would encourage Sorensen to take a vacation, but as soon as Sorensen left, Mr. Ford would start raising hell until he came back.

I learned from that, and would never take a vacation unless Mr. Ford did.

When dealing with Sorensen, Mr. Ford found it was necessary only to infer what he wanted. "I wonder what that fellow is doing here," Mr. Ford would remark about someone to Sorensen. And the next day "that fellow" was gone.

In my dealings with Mr. Ford, however, I always refused to take a hint. I'd make him tell me. But even so, after I'd carried out his orders Mr. Ford would never take the responsibility. When people would complain about something I was doing, he'd throw up his hands and say, "I don't know a thing about it."

This question of how Mr. Ford evaded responsibility is an important one, and deserves some careful explanation. Both Sorensen and I took the rap for many things that were really Mr. Ford's doings, not out of loyalty to Mr. Ford, but because we had no choice. You couldn't pin anything back on Mr. Ford.

Mr. Ford might come to me and say of some executive, "Now, you get rid of Joe. I don't want him around here any more."

I'd call Joe in and say, "Joe, Mr. Ford doesn't want you here any more. He has asked me to fire you."

Naturally, Joe didn't like being fired, and didn't mean to be, if he could avoid it. If he was up high enough in the organization, he would get to Mr. Ford and demand to know why he was being let out. Mr. Ford would then throw his hands up in that characteristic gesture and say, "I don't know a thing about it. You go back and see Bennett."

The man would then come back to me and say that Mr. Ford never told me to fire him at all. I then had no choice. *I* knew that Mr. Ford expected me to discharge the man. So I would say "I don't care what Mr. Ford told you, you're fired."

Once you've had a few experiences like that, of course you stopped saying, "Mr. Ford told me to do this." You knew he wouldn't back you up. So when he gave you an order, you just went ahead and did it and took the responsibility yourself.

If someone complained to him after you'd done what he told you to do, of course he denied all knowledge of the affair. But whatever you'd done, he'd just let the thing stand.

This explanation may seem harsh, and I dislike having to make it, but it's true. That's the way things were, for both Sorensen and myself.

I always had great respect for Sorensen, and stories that we were enemies were totally untrue. In my early years with Mr. Ford I was with Sorensen a great deal, and had no feelings for him other than admiration.

What came between Sorensen and me was not any bad feeling, but Mr. Ford's jealousy and suspiciousness.

Mr. Ford began to resent my spending a lot of time with Sorensen, and made his feelings plain. He would call me up at night sometimes when he knew I'd seen Sorensen and ask, "What did Charley want?" He openly tried to foster hostility between us. Once he said to me, "You be careful, Harry, Sorensen is no friend of yours."

After about 1930, if I went out on the floor to see Sorensen Mr. Ford made life miserable for me. Things reached a point where I had to stop seeing Sorensen except on those occasions when it was plainly necessary.

However, I didn't have much time to think about it, because I was pretty busy. Around 1921 I made my first two big moves within the Ford Motor Company.

Chapter 5

In *the early 1920s the Ford Motor Company was considered the greatest industrial enterprise in the world. When he was already forty, in 1903, Henry Ford organized the company with a total cash capital of $28,000. From 1903 to 1926 the company earned total profits reliably estimated at $900,839,000—not far short of a billion dollars. The River Rouge plant, an industrial wonder, covered 1,100 acres, or almost two square miles, and employed over 100,000 men. The company operated thirty-five branches in the United States, of which thirty-one were assembly plants. It owned vast timber lands, mines, subsidiary manufacturing plants, and a six-million-acre rubber-producing tract in*

Brazil, and operated foreign branches and associated companies in Belgium, Argentina, France, Denmark, Cuba, England, Uruguay, Holland, Chile, Brazil, Sweden, Italy, and the Irish Free State. In 1923 it was said that the company employed approximately 165,000 men in the United States and 8,000 abroad, exclusive of Canada. It has been estimated that approximately 500,000 men were indirectly dependent on the company for employment.

In this same year, just to flex its industrial muscles, the company performed a remarkable feat. On a Monday a certain load of ore was delivered at the River Rouge docks; it was cast, machined, assembled as a unit, shipped to a branch three hundred miles distant to be assembled into a finished car, sold to a dealer, and sold by him to a customer—by Thursday night.

But Henry Ford was known in these years not only for his industrial genius, but for his experiments with people as well.

In 1914 the Ford work force was 70% foreign-born, from over twenty different nationalities. The foreign worker was required to take English classes in order to qualify for profit-sharing. By this means foreign-language difficulties were greatly reduced. The Ford Motor Company took a direct interest in the health of its workers.

Expert advice was given to employees on health, housing, cleanliness—and the necessity of saving money . . .

The Sociology Department of the Ford Motor Company has been written about a great deal, and was rather famous in its time. It was set up before I came there—around 1914, I believe. It was first headed by John R. Lee, and then by the Very Reverend Samuel S. Marquis, dean of the Episcopal Cathedral of Detroit. It was a kind of "moral uplift" organization, which was supposed to help employees.

Under Dean Marquis the Sociology Department had a staff of investigators who visited every employee's home. These investigators went into every phase of the employee's private life. They asked the wives how much money their husbands saved, how much they brought home, whether they drank, and whether they had any domestic difficulties. If a workman had kept out of his pay a few dollars for a crap game or a glass of beer, he was in trouble. If he wanted to leave his wife, or if she left him, he was just about washed up with the company. The investigators became collectors of suspicions and rumors, and a card-catalogue record was kept on just how every employee was behaving himself at home. Of course, most workmen in the plant were determined to give up neither their jobs nor their crap games nor their beer, and most of them worked out ruses that enabled them to live as they wanted.

I felt the whole setup meant a stupid waste of time and money for the company and petty tyranny over the employees. If I had been one of those checked on, I certainly wouldn't have taken it. I criticized the whole thing to Mr. Ford, and he said, "Well, go ahead and stop it." So in 1921 I ended the Sociology setup as it existed, and Dean Marquis left the company.

Sorensen had nothing to do with the Sociology Department, and had no sympathy with it. He was at the tractor plant at Dearborn with Mr. Ford then, just ready to move in and take over, and was in sympathy with every move I made. Consequently, the change was an easy one to make.

This was my first big move in the company. My second was in connection with the Ford Service Department.

"Ford Service" was the name used for the plant police. Their job was to guard the gates, protect the plant, prevent theft, and keep order. But besides these ordinary police duties, Servicemen were used to check on the men constantly, to see that they kept working and broke none of the rules. How thorough a job they did is indicated by the fact that employees were even followed to the toilets.

This was going on when I came to the

Ford Motor Company. Newspapers used to try to pin Ford Service on me. But the truth is that it was an *organization policy*, and was carried out *by the organization*.

In fact, the closest you could come to pinning the policy of Ford Service on any one man would be to hand the credit to Peter Edward Martin, vice-president of the company. "Service" was his baby from the start. He was a kind of special Service Department all by himself, as long as he was there. He delighted in going around the plant to catch employees in some infraction of the rules. I have seen him snatch a cigarette out of the mouth of an engineer and then take the man's badge off and leave the train standing there before the plant with no one to run it.

It seemed to me, by 1921, that about every fifth employee was a Serviceman. Everyone was checking on everyone else. I thought the whole thing had got out of hand and was ridiculous, and I told Mr. Ford so. He said, "Well, put 'em to work," and ordered me to cut down Service by two thirds.

I did this. It was at that time that the newspapers first connected me with Ford Service. Actually, I was never the head of Ford Service. As I have said, it would have been physically impossible for me to hold down any one job, if I had wanted to.

A few years after this, our policy on

thefts underwent a change, when a Serviceman came back from a visit to the home of an employee charged with theft and boasted of snatching a Ford towel off a baby who was wearing it for a diaper.

I was so angry I wanted to quit right there, and I had a devil of a row with Peter Martin. I told Mr. Ford about it, and he was as disgusted as I was. He told me to send a whole new set of diapers to the baby, which I did.

After this incident, Mr. Ford wanted to prevent theft of tools and parts rather than apprehend people. In short, he wanted to try to lock the barn door *before* the horse was stolen. He refused to blame the workmen who pilfered something, but instead blamed the foreman who let the material get out of his control.

While this system worked moderately well in regard to the workmen, it was not so successful with the foremen and executives. When there was an excess of anything, these people usually helped themselves.

As an example of Mr. Ford's attitude on theft, I recall an incident that particularly infuriated me at the time it happened.

We had learned that a man had stolen a new motor out of the plant and installed it in his car. I told Mr. Ford about this, and he said, "Yeah? You've got the goods on him?"

I took him out to the parking lot and lifted up the hood on the man's car. Mr. Ford peered in and said, "Yep, that's a new motor, all right."

Then Mr. Ford said, "Well, by God, you just tell him he better bring his old motor in here, or there's going to be trouble."

I was flabbergasted and furious, after the trouble we'd gone to, and refused to do any such thing. But Mr. Ford himself ordered the thief to bring back the old motor.

That's the way he was. He'd use theft as an excuse for firing an executive he wanted to get rid of for other reasons, but otherwise he seemed to regard thefts as pretty petty affairs.

Tom Kelly was first head of Ford Service. Kelly was a good man, shrewd and close-mouthed, and could have run Service with many less men—if that had been the company's policy.

Kelly was discharged because Mr. Ford didn't like him and Martin wouldn't take the trouble to put in a good word for him. Here's the story.

Kelly had some of his men guarding Edsel's residence. Whenever Mr. Ford was around there, he tried to get information out of them as to Edsel's doings. But the guards weren't talking; they couldn't tell him a thing. From this Mr. Ford concluded that Kelly was on Edsel's team, and at once wanted to get rid of him.

Mr. Ford asked me to fire Kelly, and I refused. So Mr. Ford said, "Well, we'll just find out what Ed Martin thinks of him." (Everyone called Martin Pete, except Mr. Ford, who always called him Ed.)

Mr. Ford called Martin on my phone. "Will you fire this man Kelly?"

"O.K., right away," Martin said.

Mr. Ford hung up and said to me, "That's what Ed Martin thinks of Kelly."

If Martin had offered any objection at all, I think Kelly would still be there.

We had many former pugilists, both boxers and wrestlers, on the Ford payroll. We spotted them around in jobs where there was likely to be trouble. It wasn't necessary for these men to assault anyone. Just the presence of one of them in a trouble spot was enough.

An example of how these men worked out at their best is the story of Elmer De Planche, a former fighter better known under his ring name of Elmer Hogan.

When Hogan came to the Ford Motor Company he was put to work first as a fireman, then as an engineer on the D.T. & I. He did hard work, and wasn't favored in any way.

After a while the employment office asked to have Hogan transferred over to them. There was a lot of scuffling and knifing going on there, and things were getting out of hand. Sorensen had a whole deskful of knives we had taken from men both on the employment line and in the foundry.

We put Hogan in the employment office, and he did well keeping order there. Hogan was strong and fearless. He worked his way up gradually, until he was put in charge of transportation and traffic. No one else had seemed able to take care of it, because truck drivers and trainmen are a tough bunch in anyone's plant.

Hogan was always loyal. Years later, when the plant was struck, he was one of the few who stayed on the job.

My relationship with Mr. Ford was, of course, a progressive thing that developed over the years.

From the first, Mr. Ford was with me frequently, but not every day. Somewhere around 1921 or 1922 he began asking me to meet him in the plant at night, often at three in the morning; he wanted to go through the plant and check on people and see that they were doing their jobs. Sometimes Ray Dahlinger, his farm manager, went along, sometimes not. This custom persisted for years, and from the time it began my time was never my own.

Sometimes when Mr. Ford asked me to meet him in the plant, I'd wait for him there just about all night, and he wouldn't show up. The next day he wouldn't apologize, but he'd say, "That's all right. They don't know when to look for us; that keeps them on their toes and guessing all the time."

Those tours through the plant worked out this way:

Mr. Ford wouldn't talk to the foremen or any of the supervisory personnel. He'd stop and talk to men working on the job. The fact that Mr. Ford had stopped to talk to him would usually blow a man's head out of all proportion, and make him easy meat for Mr. Ford. Mr. Ford wanted information; if the man had any gripes or beefs, Mr. Ford wanted to know about them. He wanted the low-down on what was going on. As often as not, the man had something to say. When they were through talking, Mr. Ford would tell the man, "Now, if you hear anything else, just call me on the phone." I got a laugh out of that, because he was about as easy to reach on the phone as the President of the United States.

Well, the man who had talked to Mr. Ford was finished. His foreman knew the fellow had talked out of school, and got rid of him.

The next time Mr. Ford came through, he made a point of looking for that fellow. The workman was, of course, gone from the job, which was just what Mr. Ford thought might happen. Mr. Ford would question the foreman, and would be told the fellow had been shifted to another department because "he's no good." Mr. Ford knew the man had been penalized for talking to him, and he'd say, "Now, you just go get that man and bring him right back here." And the foreman would have to do it, while Mr. Ford waited.

Nevertheless, the fellow was done for. His foreman never forgot his grudge. Sooner or later Mr. Ford would forget about the workman who had talked, and then the foreman would get him out of the plant.

The tours, of course, caused consternation among the supervisory employees. Whenever they could, foremen and executives would try to warn their men that Mr. Ford was coming down the line so they'd all get on their toes and keep busy. But that was just what Mr. Ford didn't want. If he caught anyone warning the workmen, he'd become furious and have him fired.

From all this, it can be seen how nothing could happen in the plant without Mr. Ford's knowledge. He was always poking into everything that went on, gathering information in so many ways that it didn't take an executive long to learn it was futile to try to hide anything from him.

Between 1919 and 1926 Henry Ford fired most of his most talented executives. Sometimes he found, to his regret, that the generous arrangements he had made at an earlier date resulted in astronomical earnings from the increasing sales volume. William Knudsen, who had built fourteen of the Ford branch factories, found himself at odds with Sorensen and went to Chevrolet.

Sorensen and Liebold now were the major figures in the Ford Company. They could not effectively counter the export of Ford know-how which was now applied to their competition. Ford held onto his principle of a single design for too long, necessitating a lay-off of tens of thousands of employees during the time when the change finally had to be made.

The next big step in the developing closeness of our relationship came in 1927. That year marked the end of the Model T and the development of the Model A. The plant was closed down for about a year while the new model was being developed.

While the plant was closed down, Mr. Ford began dropping in to see me every day. He watched over me like a hawk. He was jealous, and didn't want me to be confidential with anyone in the plant. He usually called me the first thing in the morning, and often drove to my home to take me to work. For twenty years he called me nearly every evening at nine-thirty; if I went out to a restaurant, I was called to the phone at that hour. All during World War II we had telephones in our cars, hooked into the Willow Run Defense Plant system; my car was number nine, his was number ten. We had a code worked out so we could talk over these phones without others being able to understand us.

Our relationship became so close that I was never to be free of it for almost twenty years.

Chapter 6

WHEN THE Administration Building at the Rouge was finished, I took an office there at Mr. Ford's request.

Mr. Ford said to me, "Get in there as fast as you can and keep your eye on the accounting department. And see that the third floor stays empty."

Mr. Ford disliked and distrusted accountants all his life. As a result of his orders to keep the third floor empty, people were jammed into the second floor like sardines. Mr. Ford knew this, but wouldn't change the situation because he thought if everyone got miserable enough, that would force cuts in the office staff.

Mr. Ford not only distrusted accountants, but told me he would never be satisfied until the office force was scattered through the plant. He didn't like "remote control"; he felt that when people were too far from the work being done, they didn't know what was going on. In a way, he was right. A number of times we demolished complete buildings, and the accounting department went right on paying taxes on them, though they no longer existed.

I will never forget the rebuilding of the Truant. The Truant was an old yacht that had formerly belonged to Senator Newberry. Mr. Ford bought it with the intention of restoring it and using it for a museum piece. The ship was brought to the Rouge and the job of handling the work on it was given to an executive named Ray Rausch. As work proceeded on the Truant, Mr. Ford kept changing his plans. By the time the job was done, he had spent about a half-million dollars on the Truant, and she had been rebuilt into a brand-new yacht of latest design.

All this work had gone on without the knowledge of Edsel or any other official. One day at lunch, shortly after the work was completed, Mr. Ford turned to some accountants who were at the table with Mr. Ford, Edsel, and myself. "Are we in the black?" he asked. "Do your books balance?"

They answered, "Oh, sure they do."

"So your books balance, do they?" Mr.

Ford said angrily. "Well, we just spent a half-million dollars rebuilding a yacht here. You're nothing but a bunch of crooks."

Then Mr. Ford got up from the table and stalked away—leaving me there, holding the bag.

After the first awful moment of silence had passed, Edsel said, "Hell, you and Father were the crooks."

I said, "If I'd told you about it, I wouldn't be here."

Edsel shook his head. "I don't know what kick Father gets out of this kind of thing."

I couldn't answer that one. I didn't know, either.

Of course, this whole incident didn't make Rausch too popular with the Grosse Pointe Fords, although he'd had no choice about accepting the assignment.

And as it turned out, the Truant came to a sad end.

At the beginning of the war, the Navy took over the Truant. When she was returned, Mr. Ford's relatives got into a dispute as to who was to get the ship. This nettled Mr. Ford, and he ordered the Truant cut up with an electric torch and scrapped.

Mr. Ford never did like the administration offices. He called the officials' offices upstairs "mahogany row," and he once said to me, in the presence of a newspaperman, "They've got some nice chairs up there, Harry—but there's an awful odor."

An office had been prepared for Mr. Ford in the Administration Building, but he never used it. A woman decorator had been engaged to "do" both Edsel's and his father's offices. When she came to Mr. Ford's, she installed a fancy marble fireplace. As soon as Mr. Ford saw this, he had the marble torn out and a brick fireplace put in instead. This enraged the decorator, who said she'd been engaged to do the job, and was going to do it her way. She had Mr. Ford's bricks ripped out and the marble put back.

Mr. Ford didn't do any more about it, but the office was never used, by him or anyone else. He kept his office in the Dearborn Engineering Laboratory. When he had business at the Rouge, he used my office. If he wanted to see someone in the organization, he'd have him come there. He met people from outside in my office, too. He once told me he liked to meet people in my office because he could always get up and walk out whenever he wanted to. He did that, too, more times than I can recall. If he got annoyed or bored with some conversation, he'd just get up and walk out, leaving me there to do what I could to soothe the outraged visitor.

My office, incidentally, was in the basement. It was there because of the characters who came to the plant, and I never would leave it. We had a lot of crackpots to deal with, and there were enough of them so that they alone might have kept me busy.

For example, there was the time a man dropped in to see me. He was a fellow who had got it into his head that Edsel ought to pay him $100,000 for something or other.

The man had been trying to see Edsel, without success, and finally got in to see me. As soon as he was in my office he leveled a .32 automatic at my stomach. I saw that the safety was on, and that he knew nothing about guns. He yelled something at me, and then, before I could make a move, he swung the gun up and pointed it into his ear and pulled the trigger. I got up, talking quietly to him, and took the gun away. He went out then, babbling, "You're my friend. You're the best friend I've ever had."

One of the commonest variety we got were women who came in claiming to be Mr. Ford's mistresses. We showed one such woman a group photograph that included Mr. Ford and Luther Burbank. Her finger stabbed at the white-whiskered Burbank, a Santa Claus of a man. "That's Henry," she said.

Probably I could go on indefinitely about

all the nuts I had to deal with, but these stories are enough to show why my office belonged where it was.

I had a back door in my office in case I wanted to avoid newspapermen or subpoenas or people I wasn't prepared to see. Also, Mr. Ford was sending for me constantly to come to Dearborn, and I needed a way to get out without offending people who were waiting in the outer office to see me. In addition, the back door was used as a private entrance for Mr. Ford.

I had a target box in my office (Mr. Ford had a similar one in his garage) and Mr. Ford and I used to have target practice together, using .32 target pistols. Mr. Ford was a dead shot; he could shoot like a fool, and always carried a gun.

Sometimes Mr. Ford got bored with shooting at the target. There was a lighting fixture on my ceiling with a metal ball about as big as a marble. During the time the office above mine was occupied by W. C. Cowling, a sales manager, Mr. Ford thought it great fun to say, "Let's wake Cowling up!" And he'd start shooting at the metal ball, usually hitting it and making it ring. This would invariably scare Cowling, and he'd leave his office until the shooting was over.

In the early 1920's Mr. Ford was getting an average of five threatening letters a week. When he rode down the street, his driver had a gun under each arm. Mr. Ford had two loaded Magnum revolvers in holsters that were built into the car, and if I rode with him, I carried a gun, too.

One "shooting" incident that happened in my office got rid of an awful bore.

There was a man named Tom Bigger, then state boxing commissioner, who usually came out to see me around lunchtime. He did that for years. One day he came out with John Haggerty, a political boss. As they stepped into my office, I had just finished cleaning and loading a .38 pistol.

Bigger came in with a foul cigar in his mouth. A little annoyed to see him, I said, "Take that cigar out of your mouth."

Bigger turned his face sideways and stuck out his jaw. His cigar bobbed up and down as he said, "Go ahead—shoot it out."

I got a quick bead on it and fired, and the cigar exploded in his face.

Bigger spat out the butt and turned to Haggerty and yelped, "Come on, John, let's get out of here! The S.O.B. is crazy!"

I've never seen Bigger since.

Of course, I not only spent a great deal of time with Mr. Ford in my office and in various parts of the company, but also drove around with him a lot, wherever he might want to go. Sometimes these expeditions had interesting results. One such trip I remember vividly, because of the effect it had on the company.

Mr. Ford and I drove to a northern Michigan city on some errand or other. After it was completed, Mr. Ford decided to drop in on a dealer and check on how he was running his business.

We drove into a Ford garage, right into the back where the shop was. Mr. Ford got out of the car and looked around. The place was filthy. Mr. Ford began asking questions of the foreman of the shop, but all he got were impertinent replies.

"Who's the boss here?" Mr. Ford asked.

"The boss is busy," the foreman snapped back.

"We'd like to leave our car," Mr. Ford said.

"Jesus Christ," the foreman exploded, "can't you see we're full up now?"

I did my damnedest to let this fellow know who Mr. Ford was, but Mr. Ford kept his eye on me so closely I didn't get a chance.

Mr. Ford said, "We're driving a Ford car."

The foreman said, "A helluva lot of people are driving a Ford car."

We then tried to get into the dealer's office to see him, and you'd have thought from the treatment we received that we were a couple of thieves. The dealer was actually sitting behind a frosted glass partition in his office, and we knew it. But we were told he was in conference, and we never did get in to see him.

At last we started back for Dearborn, and Mr. Ford was furious. He told me, "By God, I'm going to get Charley to shake all these dealers up. They're all out playing golf most of the time, they won't service our cars, and they're dirty pigs. I'm going to have Charley put another agency across the road from every agency in the United States."

And he meant it.

First, as soon as we returned to Dearborn, Mr. Ford had every frosted glass partition but mine in the Ford Motor Company torn out.

Then Mr. Ford began carrying out his threat. Two executives, Harry Mack and Claude Nelis, were chosen to do the job. They started out on the road, doing just what Mr. Ford and Sorensen told them to do—set up a rival agency across from every agency in the country. By the time they got halfway across the country, all hell broke loose among our agencies. It also broke loose in the company.

Mack and Nelis had already been given their orders by Sorensen, who didn't, for reasons already explained, say, "Mr. Ford told me to do this." Edsel became enraged by the whole affair, and actually threatened to leave the company. He became embroiled in a quarrel with Sorensen, saying that Sorensen was interfering in sales. Edsel was backed up by John Crawford, his assistant, who, if anything, beefed even louder.

The main operation of the Ford Motor Company by now had been moved from Highland Park to the Rouge, and the executives were all in the Administration Building at the Rouge. The dispute between Edsel and Sorensen became so violent that Sorensen packed up and moved his whole office staff from the Administration Building right into the plant, to a spot over Gate 10 at the end of the famous overpass on Miller Road.

In the end, this worked out to Sorensen's advantage. As I've said, Mr. Ford would seldom go upstairs in the Administration Building to see anyone, even Edsel. If he wanted to see someone, he had him come to my office. However, once Sorensen moved into the plant, Mr. Ford became almost a daily visitor to Sorensen's office.

Finally Edsel came down to see me about Mack's and Nelis's operations. He came to me on this and other occasions to ask me to intercede with Mr. Ford because he simply refused to argue with his father, but he knew that I would.

By now he had found out that Mr. Ford was behind the whole thing, and he wanted the story on how his father had got off on such a rampage. I told him. He said,

"Hell, we won't have a dealer left if this keeps up. I can't even face people, the whole thing is so silly and unfair."

I agreed with Edsel, and said I'd see if I could help.

I thought that talking with Sorensen might do some good, so I went to see him first. But Sorensen only said, "Let 'em get off their fat butts."

Of course, most agencies had been working hard all along, and all were paying for one man's mistake. I finally went to see Mr. Ford, and just asked bluntly if I couldn't call those men in off the road and stop them from setting up rival agencies.

"Yes," Mr. Ford said, "only I don't think they did any harm."

"No," I said, "only to the poor suckers with their life savings in an agency."

There must be a lot of old-time dealers in circulation who can verify all this.

In 1923 Liebold began promoting Mr. Ford for President of the United States. He formed Ford-for-President clubs, and in many quarters the whole thing was taken rather seriously.

Actually, I don't know just how seriously Mr. Ford himself took this Ford-for-President thing. There were times when he joked about it. But again, there were times when he would sit in the car with me and have a vision. He would dream of making over the United States in his own way: no jails, schools everywhere, water-powered industry decentralized all over the country, and himself up there at the top, running it all.

Among other things, Henry Ford was a bird lover. He took annual holidays with John Burroughs, the naturalist. They were sometimes joined by Harvey Firestone and Thomas Edison. Both Edison and Burroughs broadened Ford's views. In turn, their appreciation of Ford's ideas was reinforcing for his independent turn-of-mind.

Chapter 7

MR. FORD began seeing me every day in the year 1927, as I've said. And in that year I performed my first big, important job for him. I settled the Aaron Sapiro case, which grew out of material published in the Dearborn Independent.

The Dearborn Independent began publishing a series of anti-Semitic articles in May 1920. These articles, which ran for ninety-one weeks, were based on the spurious "Protocols of Zion," which were exposed as a forgery in London in 1921, written by a czarist agent named Serge Nilus in 1905 in order to divert a wave of social unrest. The articles were reprinted by the Dearborn Publishing Company in four small volumes, each with its own subtitle, and all under the general title of "The International Jew." These were printed in lots of 200,000 at a time, and distributed and sold all over the world.

Of the articles in the Dearborn Independent, twenty appeared under the title "Jewish Exploitation of Farmer Organizations." It was alleged that Aaron Sapiro, a young Chicago attorney and organizer of farm co-operatives, was leader of a "Jewish ring" trying to gain control of American agriculture.

Sapiro entered a libel suit against Ford for a million dollars' damages. The case came to trial in the United States District Court in Detroit in March of 1927. Few trials before or since have claimed so much public attention.

In my early years with Mr. Ford, I never saw any bigotry in him. I don't believe anyone else ever did, either, up until around 1920. But then some of those who were close to him began to poison his mind. I believe they convinced him that some difficulties he had with a bank loan he'd made was a "Jewish plot," and went on from there to color and prejudice his every thought. Mr. Ford, after all, despite his great industrial success, had little formal education. It is my conviction that, had Mr. Ford been a better educated man, and had he not had certain people around him, bigotry never would have entered his life.

However, Mr. Ford did become a bigot.

He became bigoted about Jews, and just as much so about Roman Catholics, though the latter aspect of his intolerance was never so well known to the public.

Mr. Ford did, of course, like many Jews and Catholics. But he always had an out. When he liked a Jew—as he did Harry Newman and Hank Greenberg and Judge Harry Keyden and others—he'd say to me, "Oh, he's mixed, he's not all Jewish." When he liked a Catholic—as he liked Frank Nolan, a company lawyer, and many others—he'd say, "Oh, he isn't a good Catholic." And as soon as one of these people stubbed his toe, then right away that was a "trait," as far as Mr. Ford was concerned.

As for himself, Mr. Ford had his own private religion. He believed strongly in reincarnation. He used to expound his views on this to me at great length. One of the "proofs" he used was this: "When the automobile was new," he'd say, "and one of them came down the road, a chicken would run straight for home—and usually get killed. But today when a car comes along, a chicken will run for the nearest side of the road. That chicken has been hit in the ass in a previous life."

There were times when Mr. Ford tried to convert me to prejudice. But I'd never had any feeling of that kind, and the training I got from my mother, who was a fine, principled woman, saved me from being susceptible. But if I had been bigoted when I met with Mr. Ford, I believe he would have cured me, when I saw the asinine things people told him, and how he swallowed their lies.

I only saw Mr. Ford ashamed of his bigotry before one man, and that was Thomas Edison. Mr. Ford almost worshiped Edison. More than once I heard Edison rebuke Mr. Ford for his prejudice. Mr. Ford always denied it to him. "The damn rotten newspapers are making me look that way," he'd say.

And now I want to talk about two men I've mentioned before, Ernest Liebold and Bill Cameron. This is a good place to do it, since Cameron was editor of the *Dearborn Independent*, and Liebold, among his other duties, was general manager of the parent company, the Dearborn Publishing Company.

During all the time I was with Mr. Ford I was completely antagonistic to both Liebold and Cameron. I made endless attempts to fire them, until, in Liebold's case, I succeeded. It is hard for me to say which one I disliked most, but I guess honors would go to Liebold.

Liebold looked like a typical Prussian, and he often acted like one. Mr. Ford once told me that when dinnertime arrived in the Liebold household, Liebold marched his children around the table in military style, and when they had reached their places he commanded, "*Sitzen sie!*" He had a Gestapo of his own within the Ford Motor Company; he kept elaborate files, and had something there about everyone.

Bill Cameron was a short, stout, round-faced man; he looked and talked a lot like W. C. Fields, with the difference that Fields was funny. I have heard that he was once a preacher in Brooklyn, Michigan. He came to the Ford Motor Company from the *Detroit Daily News.* I never could like Cameron. To me there was something fundamentally unpleasant and irritating about him. As many times as I tried to sit down and talk with him in a friendly manner, I always ended up angry.

Cameron and I were enemies almost from the very beginning. Back in the early days when Cameron was very close to Mr. Ford, and I had but little standing in the company, I slapped Cameron's face in my office for using profanity before a young woman. He took it, too. Backing out of the room, he said, "By God, I didn't think you had the nerve."

For the thirty years that I knew him, Bill Cameron was quite a drinker. When he became the commentator on the Ford Sunday Evening hour in 1934, two men were assigned to the job of getting him to the studio.

Mr. Ford, inconsistent in so many things, was also inconsistent in his hatred of drinking. He might fire a workman in the plant caught with liquor on his breath, but when it came to someone like Cameron, his attitude was different. I don't know how many times, at Mr. Ford's request, I had to help Cameron—only to have him hate me for it. After a while, our mutual hostility grew so that Cameron refused to talk to me in person, and if I called him on the phone, he just hung up.

Both Liebold and Cameron, but particularly Cameron, were constantly stirring up Mr. Ford. I recall in particular one time when Liebold sat beside Mr. Ford at a big banquet. I saw Liebold lean over toward Mr. Ford and heard him whisper, "See that man over there? He's a Jew." And then I saw Mr. Ford look at the man Liebold had pointed out, and his expression changed.

Well, to get back to the Sapiro trial. From the very beginning, I urged Mr. Ford to try to settle things. I wanted him to let the organization get back to building cars.

Mr. Bennett's concern was justified. The Model T had come to the end of the line. Henry Ford had produced fifteen million of them at an average rate of 1.6 per minute, and had sold them in every corner of the globe for an aggregate sales gross of seven billion dollars. But by 1927 the public wanted more style and comfort than the T afforded, and in that year, for the first time in its history, the Ford Motor Company lost money.

But my advice did not prevail until some strange things had happened.

The Aaron Sapiro trial took place in Detroit's Post Office Building in the court of Federal Judge Fred M. Raymond. Mr. Ford was represented by a legal staff of seven attorneys, headed by Senator James M. Reed of Missouri. Mr. Ford expected that Sapiro would be represented by a "Jew lawyer" from New York, and thought this would reflect to his own advantage with the jury. However, Sapiro walked into court represented by Attorney William Henry Gallagher, an Irish Catholic. Mr. Ford considered Gallagher a "Christian front" for Sapiro, and after that always spoke of the Catholics as "tools of the Jews." The trial began on March 15, 1927.

Before the trial began, Gallagher told the press that a subpoena had been served on Henry Ford at the Ford Airport on August 6. In answer to this, C. B. Longely, general counsel of the Ford Motor Company, informed the press that his client had given no intimation that he had been summoned.

At a hearing preliminary to the trial, held on March 14, Gallagher asked Judge Raymond whether Henry Ford was going to answer the subpoena and appear in court the next day; for if Ford were not going to appear, as intimated by Longely in the press, then Gallagher wanted time to prepare contempt papers. Judge Raymond referred this question to Ford attorneys, and Hanley of Ford counsel replied that he did not know his client's intentions.

On the first day of the trial I took Mr. Ford to the Post Office Building and up to his lawyers' quarters. We sent word to Sapiro that Mr. Ford was ready to testify whenever he was wanted. Sapiro's attorneys and Mr. Ford's attorneys had a conference, and by agreement fixed the following Monday as the date for Mr. Ford's testimony.

A jury of six men and six women was selected that first day. As the trial itself got under way, two different struggles developed between opposing attorneys.

The first was over the question of what was admissible evidence. Mr. Ford's attorneys wanted to keep the trial strictly on the question of whether or not Aaron Sapiro as an individual had been libeled. Sapiro and his attorneys wanted to widen the question; they wanted to include the question of libel of the whole Jewish people. I believe Sapiro lost out on this, and the court refused to permit Mr. Ford's attitude toward the Jewish people to become part of the record.

The second struggle that went on was over the matter of Mr. Ford's testimony.

Sapiro had subpoenaed Mr. Ford, but I

think he wanted to put him on the stand only as a last resort. If Sapiro had called Mr. Ford as his witness, he could not have cross-examined him or brought in other witnesses to refute his testimony, or even questioned any statement Mr. Ford chose to make. This would have been a very advantageous situation for Mr. Ford.

Sapiro must have hoped that Mr. Ford would take the stand on his own behalf. If he did that, then Sapiro and Gallagher could cross-examine him at will and bring in witnesses to refute his testimony. I suppose it was as a result of this hope that Sapiro's attorneys kept stalling on calling Mr. Ford as their witness.

As for Mr. Ford's attorneys, they wanted to have Sapiro call Mr. Ford to the stand. They saw that such a situation would be to his advantage. They also thought that Mr. Ford's public prestige was such that a personal appearance in court by him would have a strong influence on the jury.

As for Mr. Ford, I know he didn't want to testify. At first he was willing, though reluctant, to testify, if Sapiro called him to the stand. But the more the lawyers stalled, the more nervous Mr. Ford became. I guess he remembered Mt. Clemens.

The trial began with Cameron as the first witness. He testified that Mr. Ford had had no knowledge of the Sapiro articles at the time they were published. Over a period of about five days on the witness stand, Cameron took all responsibility for everything that had ever appeared in the *Dearborn Independent,* and said, in effect, that Mr. Ford had no connection whatsoever with the editorial policy of the paper. He testified, "I run the paper and use my own judgment."

I don't know about that. But during the time Cameron was speaking of, Mr. Ford dropped into Cameron's office just about every day of the week.

After Cameron had stated at length that the Independent *represented only his own views, and no one was going to tell him how to run it, Gallagher, with a smile, asked the witness if his own views on the prohibition amendment coincided with those of the* Dearborn Independent.

The question was objected to by Ford counsel, and sustained.

When Cameron began his testimony, Sapiro and his lawyers announced that Mr. Ford would not be called to the stand at the previously agreed date, but that Sapiro would follow Cameron to the stand. Somewhere in there I took Mr. Ford down to the Post Office Building, but was told that the lawyers were not ready for him. With this second postponement (Mr. Ford had expected to testify the first day of the trial), Mr. Ford's nerve began to fail visibly.

When Cameron's testimony was finished,

he disappeared somewhere in Canada. It took us days to find him.

Sapiro took the stand as the next witness, and again I took Mr. Ford down to the Post Office Building. I left Mr. Ford sitting in the car and went up to the lawyers' quarters myself.

Longely said, "Oh, God, we can't use him now."

The lawyers' quarters smelled of whisky. I knew that if I brought Mr. Ford into that atmosphere he would do something rash. I didn't want Mr. Ford to go in there, and I went back to the car and told him so.

"Well, Harry," Mr. Ford said, "I want to stop this. I'm not coming down here again."

Two days later the newspaper called me at my home. They said Mr. Ford had been in an automobile accident, and what did I know about it? I said I didn't know a thing about it, which was the truth, and tried to stall them along. They told me the story—embodied in a formal statement issued by Cameron—which was then on their presses.

It was said that the previous Sunday Mr. Ford had been driving alone in a Ford coupé from the Dearborn laboratories to his residence; that a big touring car driven by two men had knocked Mr. Ford's car off a bridge crossing the Rouge River. It was stated further that after a period of unconsciousness, Mr. Ford had walked to his gatehouse in great pain, and the gatekeeper called Mrs. Ford, who helped Mr. Ford into the house and summoned his physician. Mr. Ford's physician had stayed with him two days, and then taken him by ambulance to the Henry Ford Hospital, where an operation had been performed. It was said the statement had been held up two days because of the "unavoidable and unfounded inferences that may be drawn" —thus neatly inferring that Sapiro and/or his agents had made an attempt to kill Mr. Ford.

As the newspapermen told me all this, I could hear the plant trying to get through to me, swearing at the reporters. Finally I got the reporters off the line. The plant told me that I should stop at the Residence on my way to work.

I went to the Residence, and there saw Mr. Ford. He looked all right to me. I said, "The papers said you have a broken rib."

"Did they?" Mr. Ford said. "Well, maybe I have."

I said, "I'm going to find out who knocked you into the river if it takes me the rest of my life."

"Now," Mr. Ford said, "you just drop this. Probably it was a bunch of kids."

I kept at it. I was half indignant and half skeptical. On my way to the Residence I had stopped at the scene of the "accident" and looked around, and there were things that seemed phony to me. I said,

"No, I'm not going to drop it. If someone has tried to kill you, I'm going to find out about it. I don't have to work for you to do that—I can do it on my own."

Finally he saw there was no way to put me off, and he said, "Well, Harry, I wasn't in that car when it went down into the river. I don't know how it got down there. But now we've got a good chance to settle this thing. We can say we want to settle it because my life is in danger."

The case dragged along for a few more weeks. Sapiro was kept on the stand by an exhaustive cross-examination from Senator Reed.

During this period I saw Mr. Ford almost every day at the Residence. We had a large body of investigators checking the courtroom and following people around to see what we could "get" on someone, thinking we might settle the case that way, but without results.

Finally, one day when I was with Mr. Ford, he gave me some information that had been brought to him, which purported to be an attempt at bribing a juror. I thought this evidence pretty slim, but I saw a chance to use it.

From the very beginning of the trial, I was persistently followed everywhere I went by a man named Hutcheson, a Washington correspondent for Hearst who was covering the trial. It seemed that everywhere I went, he popped up.

I now said to Mr. Ford, "Do you want to settle this thing? If you do, I'll give your tip to this fellow Hutcheson who's been following me around. He'll print it, and the judge will toss the jury out. Then they can settle it."

Mr. Ford told me to go ahead.

By this time, Gallagher had announced that he intended to demand an examination of Mr. Ford by an impartial physician if Mr. Ford did not come to court when called.

I approached Hutcheson and told him about Mr. Ford's information. "Mind," I said, "this isn't something I can prove. It's just something we've heard."

"Just tell me what it is," Hutcheson said, "and damn right I'll print it."

Our lawyers then took Mr. Ford's allegations up with the judge. They gave him fourteen affidavits alleging various irregularities. The judge turned this information over to the FBI for investigation. Sapiro was not informed of this development. Judge Raymond called in all the newspapermen covering the case and warned them to print nothing about the matter. Hutcheson was conspicuous by his absence.

Hutcheson wrote a story based on the affidavits that our attorneys had filed with the judge, and turned it in to the Detroit

Times. They printed the story under screaming headlines.

When the *Times* appeared on the street, Judge Raymond at once said that the story constituted contempt of court, and that he would start proceedings against the paper.

Mr. Ford's attorneys now filed an application for a mistrial with Judge Raymond. Judge Raymond granted the mistrial the next day, April 21. He completely exonerated Sapiro of any charges of jury tampering.

A few months passed, and before the case could come up for retrial, Mr. Ford settled it out of court. Not much of that story is known.

The name of Herman Bernstein has never been publicly mentioned in connection with the Aaron Sapiro case, but he had more to do with Mr. Ford's repudiation of anti-Semitism than anyone else.

Herman Bernstein, editor of the Jewish *Tribune,* went to Europe in 1915 with Mr. Ford on the "Peace Ship." Years later, in the *Dearborn Independent,* Mr. Ford claimed that Bernstein had told him on the "Peace Ship" that the "international Jew" was responsible for World War I. Bernstein vigorously denied ever having said this.

Well, Bernstein came to see Mr. Ford shortly before the Aaron Sapiro trial began. They had a long and bitter discussion about Mr. Ford's bigotry, and Mr. Ford claimed that nothing he had ever caused to be printed had hurt anyone. Bernstein insisted that Mr. Ford was wrong, that "The International Jew" had actually stimulated real physical violence against Jews in Europe. "If you can prove that," Mr. Ford said, "I'll take back everything I've ever said."

Bernstein promptly departed for Europe. He made a five-month tour of Europe and Asia, returning to New York on June 9, 1927, after the mistrial had been granted. He brought to Mr. Ford documentary evidence that Mr. Ford's *Dearborn Independent* had indeed hurt a great many people. When he saw this evidence, Mr. Ford was ready to quit publishing anti-Semitic material.

Mr. Ford sent me to New York to settle the case.

I got in touch with Arthur Brisbane, and through him learned that the American Jewish Committee could settle the matter. I entered into negotiations with Samuel Untermeyer and Louis Marshall, of that organization, and with Brisbane.

Throughout this case, Joe Palma, then of the United States Secret Service, later a Ford dealer, did a lot of work to help me. He was my one confidant in New York, and my intermediary with Brisbane, Marshall, and Untermeyer. Palma, incidentally, was deeply hated by Bill Cameron, who did everything to dislodge Joe from the Ford Motor Company; but Mr. Ford stuck with me in keeping Palma with the organization.

Brisbane, Untermeyer, and Marshall drew up the now-famous "apology," which was to be the basis for a settlement. In this formal statement, it was said that Mr. Ford would see to it that no more anti-Semitic material circulated in his name, and that he would call in all undistributed copies of "The International Jew." For the rest, the "apology" said that Mr. Ford had had no knowledge of what had been published in the *Dearborn Independent,* and was "shocked" and "mortified" to learn about it.

Arthur Brisbane brought this statement to me at 1710 Broadway. In the presence of Gaston Plantiff, I called Mr. Ford. I told him an "apology" had been drawn up, and added, "'It's pretty bad, Mr. Ford."

"I don't care how bad it is," Mr. Ford said, "you sign it and settle the thing up."

I tried to read the statement to him over the phone, but he stopped me, saying again, "I don't care how bad it is, you just settle it up." And he added, "The worse they make it, the better."

So, in the presence of Gaston Plantiff, I signed Mr. Ford's signature to the document. I had always been able to sign his name as realistically as he could himself. I sent the statement to Untermeyer and Marshall. The signature was verified, and the case was closed.

All this was done without Mr. Ford's taking anyone else into his confidence. Edsel knew nothing about it, and Cameron and Senator Reed heard about it by reading the papers.

It is said that Senator Reed telephoned the Ford plant from Texas and demanded, "What in hell is this I see in the Dallas paper?" And Cameron was quoted by the newspapers: ". . . It's all news to me and I cannot believe it is true."

Mr. Ford paid Aaron Sapiro's legal expenses, and he also reimbursed Bernstein for the expenses incurred on his trip. Neither man would take a cent over that.

The "apology" was printed in the *Dearborn Independent,* and the paper ceased publication early in 1928.

Chapter 8

WHEN the Sapiro case was settled, leading Jews of Detroit were very much put out because Mr. Ford had gone to New York to settle the case, rather than going to them. In particular, Mr. Ford exchanged some pretty harsh words with Leo Butzel, a prominent Detroiter. As a result of this, Mr. Ford got a kind of persecution complex on Leo Butzel.

At one time, Mr. Ford wanted to know what Leo thought of him, and asked me to find out. I sent a man named Taylor to Butzel's summer house at Mackinac Island, where Taylor got a job as a gardener.

Now, it happened that Taylor was a fine pianist. Leo found out about this and began inviting Taylor into the house to play for him. As a result, the two men became fast friends.

Finally, Taylor came back and wrote up a two-page report on Leo. He said, in a glowing eulogy, that Leo Butzel was the most wonderful man he'd ever met.

I gave the report to Mr. Ford, and he was furious. "The man's a liar," he said, and ordered me to transfer Taylor from Dearborn to the Rouge, so he wouldn't have to see him again.

But that wasn't the end of it.

At about this time Charles Bowles, mayor of Detroit, was recalled, and Frank Murphy was elected mayor in his place.

As the recall of Bowles got under way, Mr. Ford decided that the whole thing was a plot by Leo Butzel, and promptly decided to help Bowles. Into this decision, too, entered the fact that everyone was down on Bowles; when everyone was down on someone, Mr. Ford was all for him.

Bowles was not a very pleasant person. Once when we had him out to lunch he ate so much that he almost died in my office. We had to carry him to a first-aid station, where a doctor pumped him out.

Nevertheless, Mr. Ford wanted to help Bowles. Mr. Ford, Frank Nolan, and I made a public call on Bowles. Mr. Ford ordered Lou Colombo, then general counsel for the company, to go into court and defend Bowles.

As all this was going on, a lawsuit involving the Lincoln Motor Company was approaching the State Supreme Court. Mr. Ford became more and more apprehensive about the outcome.

As a result of this apprehension, he began to worry about State Supreme Court Justice Henry Butzel, who was Leo's cousin. Mr. Ford didn't have anything against him—except that he was Leo's cousin. Mr. Ford thought that because of this relationship, Justice Butzel would be inclined to rule against him.

Butzel had been appointed to the bench by Governor Fred Green, and shortly before the case over the Lincoln Company came up before the State Supreme Court, he was up for election. Mr. Ford offered me any amount of money I wanted if I could defeat Butzel.

I was upset by this determination on Mr. Ford's part. I went to see Governor Green and told him that Mr. Ford wanted me to try to defeat Butzel. When I told him this, Fred hit the ceiling. He said Butzel was one of the outstanding citizens of Michigan, and that he didn't mean to see him defeated. He insisted upon seeing Mr. Ford to talk with him about the matter, and I took him to Mr. Ford's office.

When we sat down to talk with Mr. Ford, he disclaimed all responsibility. Mr. Ford turned to me and said, "I'm not interested in the Supreme Court. Once someone is appointed to that court, whatever he was before, he becomes a fair man."

This satisfied Governor Green, and he left. But as soon as he was gone, Mr. Ford said to me, "Harry, I guess this ends our confidential talks together."

The next day I didn't come to the plant.

At about four-thirty in the afternoon, Mr. Ford came out to my home, driven by Ray Dahlinger. He was entirely changed in his attitude toward Justice Butzel. He told me, "I was just mad, and I said things I didn't mean."

Mr. Ford never mentioned Justice Butzel to me again.

Sales managers came and departed one after the other at the Ford Motor Company. We'd get a new sales manager, a new model would come out, and he'd go along fine for a while. Then sales would slump, and inevitably the sales manager would ask for changes in the car. When he criticized the car, the sales manager was through.

I had nothing to do with any of these changes, but it was always my unpleasant duty to carry the fatal message to either Sorensen or Edsel.

As I've said, the Model T was discontinued in May 1927, shortly after the Sapiro trial began. The plant was closed down for about a year, and during that time the Model A was developed and retooling carried out. The "A" had many radical changes over the "T," and was a first-rate car.

Although tardy in recognizing the realities of the marketplace, Henry Ford incorporated in the Model A all the variations that he had previously opposed. The Model A had as standard equipment automatic windshield wipers, gas and oil gauges, a foot throttle, a regular gear shift, shock absorbers, four wheel brakes, and was made available in four different colors.

Henry Ford's reputation as a manufacturer was such that 500,000 customers made a down payment on the Model A before they knew its price and before they had ever seen it. Madison Square Garden in New York City was a pandemonium when the car was exhibited there; Cleveland mounted police were called to keep mobs from breaking in Ford show windows in that city; and 25,000 people turned out in subzero weather in St. Paul to see the Model A.

As the Model T went out, so did William Ryan, our sales manager, who had succeeded Norval Hawkins. He had been with the company in one capacity or another for about nineteen years.

Bill Ryan was the only sales manager who, both Mr. Ford and Edsel agreed, was a perfect sales manager. "We'll never get another sales manager like Ryan," Mr.

Ford used to say. Nevertheless, it was Mr. Ford who let him go. Why, I am not sure.

Fred Rockleman, who followed Ryan and came in with the "A," lasted only three years as sales manager. Rockleman had started with the Ford Motor Company in 1903 as a mechanic. He had worked his way up to manager of a branch plant, and later was put in charge of the D.T. & I., where he did a distinguished job. I guess Mr. Ford thought he could do almost anything well.

There may have been contradictory elements in Henry Ford, but in one thing at least he was consistent. The success of the Model A resulted in a freeze on engineering development and the people's motor car remained unchanged for five years. Meanwhile Knudsen at Chevrolet had a six-cylinder motor competing with the Model A's four cylinders. Chrysler moved in with a very appealing Plymouth.

After a few years, as sales manager on the "A," Rockleman made that familiar mistake—he criticized the car. He said the radiators were overheating. They were, too.

Rockleman wanted hydraulic brakes on the car, and other improvements made, to meet competition. He took these things up with Edsel—because you could reason with Edsel about style, engineering, and sales, but not Mr. Ford; the only thing he cared for was what was under the hood. But when Rockleman began taking up everything with Edsel, that was his undoing.

As sales continued to slump on the "A," Edsel began to favor changes in the car. Also, he began wondering, out loud, why the Ford Motor Company didn't have men like General Motors' Kettering, etc. The men Edsel mentioned in this vein, of course, at once went down on Mr. Ford's crap list. Mr. Ford felt that Edsel was being influenced by Rockleman, and was burned up about it.

This is a classic example of how, when Mr. Ford felt someone was influencing Edsel against Mr. Ford's own ideas, he came down on the offender like a ton of bricks. Once Mr. Ford became convinced that Rockleman was turning Edsel away from his own ideas, Rockleman never had a chance.

I took Rockleman's final check over to him. He laid all his troubles to Sorensen, but he was wrong. Mr. Ford had kept at Sorensen relentlessly to get rid of Rockleman. Only, again, here was another case where one couldn't say, "Mr. Ford told me to do this."

W. C. Cowling followed Rockleman as sales manager. Also an old-time employee, Cowling lasted about seven years.

By the time Cowling left, Mr. Ford had been crabbing about him for years. Edsel didn't like him, either. The truth of the matter is that Cowling talked himself into this job, and then talked himself out. He was a great orator—one of the best we ever had. But that was not an asset, because Mr. Ford, who always spoke in telegrams himself, disliked orators.

In my own opinion, Cowling was put in the wrong job. He was really a top-notch traffic man, and should have stayed there. He never should have been put in sales in the first place.

Cowling, of course, also made that fatal mistake of criticizing the car. In addition to all this, Mr. Ford became angry and suspicious because he heard that Cowling was building a fine new home.

Cowling's two sons had an interest in some shipping companies that did business with the Ford Motor Company. Perhaps Mr. Ford felt Cowling was getting additional income from this source, and had his usual reaction to executives having outside interests.

Mr. Ford decided to try to force Cowling to resign, and for this purpose laid a plot.

For quite some while, in order to stimulate sales, the Ford Motor Company had a policy whereby cars could be reconditioned by the factory very cheaply. All the executives took advantage of this. We'd bring our cars in, get a new chassis, a new engine, pay a small bill, and drive away an almost new car.

Cowling did this, too. But Mr. Ford was laying for him. He told me to keep track of all the time that was actually spent on the car, and then add to that a charge for the time he and I spent checking on the job, and send a bill to Cowling.

When the car was finished, we sent Cowling an enormous bill. But it didn't have the desired effect. He called me up and said he'd got the bill, and then added, "You can keep the so-and-so car."

However, I guess he got the idea, because shortly after that he told me he had sold the home he was building.

Finally Mr. Ford and Edsel made up their minds about Cowling. Edsel brought Cowling's check to me and said, "Let me get out of town before you give it to him."

Cowling was followed by Jack Davis.

Like all the rest, Davis went great guns with a new model. Then sales began slipping. Davis went out to the West Coast on a business trip, and from there wrote back a letter saying the car was no good, and that's why it wasn't selling. Mr. Ford and Sorensen felt Davis's comments were purely destructive criticism, and were enraged.

At about this time Davis got into some domestic difficulties, and Mr. Ford seized on this as a pretext and ordered me to fire him.

Shortly after I got this order, Edsel called me and asked me not to do it.

Of course, this put me in a fine spot. But I liked Davis, myself, so instead of firing him I transferred him out to the West Coast, without telling Mr. Ford. Eventually, of course, he found out about it anyway, but by then he'd cooled off.

Next on the list was Clay Doss.

Doss had been with the company many years earlier, had quit, and had come back. He became our branch manager in Kansas City.

When Mr. Ford and I attended the American Legion convention in Chicago in 1939, Doss had by then left Kansas City and become our Chicago branch manager. Doss took care of us and our retinue, and did this so well that Mr. Ford came back to Detroit sold on Doss. He thought we ought to give Doss something bigger to do.

Edsel grumbled a bit, reminding his father that Doss had quit once, and then suggested that we send Doss to New York as sales head there, and bring Cap Edmund, by then sales head in New York, to Detroit as general sales manager. But Mr. Ford said, "No, give that job to Doss."

So Doss came to Detroit as sales manager.

Shortly after this, young Henry, Edsel's son, came into the plant. Doss at once decided to take over young Henry. Doss told someone he wanted to become a strong influence on Henry. He also made a certain remark that, when Henry heard about it, he resented very much.

Doss and Henry took a trip to Washington together, and whatever took place, Henry came back with a lungful of disapproval of Doss. Then Doss left the company.

When this happened, Mr. Ford said to me, "What's the use of having any more sales managers? We'll just let them go."

I then suggested the zone-manager system, in which sales authority is considerably decentralized through the various regional areas. Jack Davis, at my suggestion, was brought back and put at the head of the zone managers. The zone-manager system is still in effect in the Ford Motor Company today, I think.

So much for the parade of the sales managers. But before I leave them, I must go back to the reign of W. C. Cowling, for an incident occurred then that will illuminate considerably the relationship between Mr. Ford and myself.

Toward the end of Cowling's time with the company I decided to buy some life insurance, and began conversations with an agent with a view toward taking out a policy. One day when the agent was in my office talking with me, Mr. Ford walked in.

I told Mr. Ford what I was doing. His reaction was one of rage. He turned on the salesman and berated him at great length for trying to sell me a policy. Then he turned to me and told me that he intended to give up his own insurance, and explained, "It's a bad thing, Harry. It makes people hang onto you because of what they'd get if you died."

Mr. Ford reminded me that Cowling's sons had an interest in some shipping companies, and said, "Now, if you're worried about your children, give them something like that."

So we gave my daughter stock in a shipping company that Joe Palma had set up in New York for the Fords.

This was supposed to be my insurance. Actually, the returns from this were so small they didn't amount to much, and I never took it very seriously as a form of security for my family. It was fortunate that I didn't. As soon as I left the company, the shipping company was canceled out.

But Mr. Ford's reaction to my plan to carry insurance throws light on his attitude toward me. I believe that Mr. Ford thought of me as a son. He was always extremely solicitous of my health, and, I think, felt as close to me as a father might. But like many another father, he wanted me dependent on him.

Time and again, situations arose where it became plain that Mr. Ford didn't want me to have much money. I got peanuts for a salary for twenty-eight of the thirty years

I was with him; the last two I got my salary up, but that is a story to be told later. On the other hand, Mr. Ford would give me almost anything I wanted as a gift. Yet that is significant, too, because to receive gifts is to be dependent on the giver.

Mr. Ford maintained my homes for me. He sent out crews to paint a house, for example, whether I wanted it painted or not, and in every way kept them up. That wasn't always an advantage, though, because with his passion for interfering in other people's lives, he usually injected his own ideas into any decorating or rebuilding that went on, and had the thing done the way he wanted it, not the way I wanted it.

Even the business of gifts didn't always work out. Mr. Ford would do a quiet good deed for you—send you a new refrigerator, for instance—and then just as quietly he'd tip off a few people that he'd done it. This would, of course, stimulate jealousy among other executives, and in general not contribute much to your popularity.

I guess what Mr. Ford wanted was to have me look for all my security in the closeness of our personal relationship. And I must admit, now, that it worked.

While I was with Mr. Ford I paid a great many expenses out of my own pocket. Sometimes I did this because Mr. Ford wouldn't pay them, and I'd be embarrassed. Sometimes I paid them so there wouldn't be any questions raised about the expenditures. There were actually occasions when I paid out my whole salary within five minutes of the time I received it. But that never bothered me any. I believed I was secure.

Chapter 9

THOUSANDS of former criminals were taken on the Ford payroll over the course of the years, at Mr. Ford's request. The hiring of such people by the Ford Motor Company began many years before I worked for Mr. Ford.

Mr. Ford gave the public to understand that he hired these people for the purpose of rehabilitating them. But I didn't see much of that sort of thing happen. I guess we helped a few of the youngest ones straighten out. Otherwise, they were a constant headache to me. There'd be a crime committed in Detroit, or a check forged; the police would decide that the whole thing looked like the handiwork of one of our employees, and they'd tear right out to the plant.

Mr. Ford often had a "practical" motive in hiring a given ex-convict, as I shall explain later. Besides, I think, he felt the "rehabilitation" angle made for good free advertising. But beneath all this, I think, his motives for hiring former criminals lay deep in his own personality.

Mr. Ford had a profound morbid interest in crime and criminals. He also had a deep sympathy with them. He used to dream of a day when there would be no jails, and he was violently opposed to capital punishment.

Whenever we hired a former criminal, Mr. Ford always wanted to come to my office and talk with him.

He'd say, "Now, how did you get into this?" And then, sometimes before the man would have a chance to answer, he'd say, "I'll bet a woman got you into it." That was his theory of crime—he always looked for a woman at the bottom of the trouble. He liked to say, "I can tell a man who has a good wife by the way he's getting along."

To show how these people came on the payroll, I shall tell the story of the hiring of Buff Ryan and Bishop Ablewhite.

Buff Ryan, a well-known Detroit gambler, was indicted and sent to Jackson Prison for bookmaking. When this happened, Mr. Ford came to see me and insisted that I find out who Ryan was protecting. However, a grand jury took on that unpleasant job and saved me the trouble.

Mr. Ford always had a sympathetic feeling for gamblers, and would never criticize one. A number of professional gamblers were his friends.

I remember one time a man came to see me, to ask how he could get around Mr. Ford in order to get horse racing into Dearborn. I spoke to Mr. Ford about it.

"Why," he said, "I'll go to see the races myself." And then he added, "Harry, I'm the biggest gambler in the United States. Here's my gamble—right here." He meant, of course, the Ford Motor Company.

Besides this feeling of sympathy, Mr. Ford's interest in gamblers was also due to the fact that they were excellent sources of information. There isn't much goes on that gamblers don't know about.

While Ryan was in prison, Mr. Ford came to me periodically to ask about him. Finally we were able to have Ryan paroled to me by the prison parole board.

Mr. Ford came to my office to meet Buff Ryan. He asked Ryan many questions about prison and prison life. After a long talk Mr. Ford got up and said to me, "You hire him, right now."

Ryan had been feeling pretty licked, and tears of gratitude came to his eyes. "I never worked in a factory in my life, Mr. Ford, but I'll do my best," he said.

Mr. Ford started for the door, then turned. "You don't have to work in the plant. Just keep your eyes and ears open. We want to know what's going on around town."

We didn't make any secret of this connection. John Bugas, formerly head of the Detroit FBI office, was informed of it, and even the newspapers knew about it. But some years later there were efforts made to have the whole arrangement appear as something done behind Mr. Ford's back.

When young Henry came into the plant, he stopped in at my office and said he thought it was a terrible thing for a man of Ryan's reputation to be on the Ford payroll. I told Henry there were men on the Ford payroll who would make Ryan look like a Sunday-school boy.

The case of Bishop Ablewhite was somewhat different. Ablewhite, an Episcopal bishop, was sent to prison for gambling with the funds of his church.

After Ablewhite had been in prison for some time, Mr. Ford asked me to get Ablewhite out and put him to work in our Sociological Department. It was a distasteful job to me. While I can sometimes sympathize with the man who goes wrong out of poverty and ignorance, I have no sympathy at all with a man who has a bishop's advantages and background, and who still goes wrong. But I had my orders, and I managed to get Ablewhite paroled.

Ablewhite came to my office to see me. I sent him to Miller, our employment manager, with instructions that he was to work in Sociological under Leonard Saks, who had been Jack Dempsey's manager.

The Bishop had been out of my office only a short time when Mr. Ford came in. And Mr. Ford was no sooner settled in a chair than my phone rang. It was Saks, and he said in his soft voice, "Boss, what do you want me to do with Abie the Bish?"

I told Mr. Ford what Saks had said, and he laughed. From that time on, Mr. Ford as well as the rest of us always referred to the man as "Abie the Bish."

Well, after Abie the Bish met Leonard Saks, he went right back to see Miller again, full of indignation. "Why," he said of Saks, "that man's a Jew. I wonder if Mr. Ford knows that."

Miller told me about the Bishop's crack, and I was burned up, and told Saks.

I guess Abie the Bish didn't have a very happy life after that. Saks sent the man out on the worst assignments he could find. When Abie the Bish came back from one of these assignments, he invariably went in to see Saks and complained about the miserable experience he'd just been through. Saks was very sympathetic; he sat there and listened, sometimes with real tears streaming down his face because of the Bishop's unfortunate experiences.

When I left the Ford Motor Company, Leonard Saks walked out. At this writing, Abie the Bish is still working in the Ford Sociological Department.

The hiring of former criminals at the Ford Motor Company was only a part of our association with the criminal element, however. We also set up contacts with the underworld. We did this for two reasons. First, we did it because Mr. Ford was deathly afraid that kidnapers would be after Edsel's children (he had four: Henry II, Benson, Josephine, and William). It was Mr. Ford's idea that he could get protection by dealing with these people. Secondly, they served as a source of information.

Only if you'd lived in Detroit during prohibition times could you know how bad the crime situation was. Detroit was, I should say, on a par with Chicago—until the newspapers began a crusade that cleaned up the city. It actually got so bad in Detroit that, rumor had it, the police asked gangsters to please dump their bodies outside the city limits.

I personally felt that Mr. Ford's fears for Edsel's children were without too much foundation; I felt that children were getting too hot for gangsters to handle, because of the public's outrage at such kidnapings. It was my conviction that Edsel himself was in far greater danger.

However, in deference to Mr. Ford's wishes, I assigned two men named Frank Holland and Jim Brady to watch over the children. They did so for a long time, and did an excellent job of guarding them against danger. However, they were not thanked for this.

With Edsel's full knowledge and consent, we also had men tailing Edsel wherever he went.

Through Bert Brown of the United States Secret Service I was introduced to the man who, it was claimed, was the head of the Detroit underworld—Chester LaMare, a short, bull-necked, swarthy Sicilian. Through Joe Palma, I also met other men of the underworld in both Detroit and New York.

We gave LaMare a Ford agency, known as the Crescent Motor Sales Company. Later we discovered he was using it as gang headquarters.

We also gave LaMare the fruit concession at the Ford Motor Company—in an indirect way.

The fruit business at that time was in the hands of gangsters—a situation that led to considerable bloodshed. When gangsters shot a little peddler through the eyes (he lived through the experience) Mr. Ford got disgusted. He told me to find out who controlled the fruit business in Detroit, and give the concession to them, with the understanding that there was to be no more violence, or we would not permit fruit to be sold at the plant.

I learned that a man named Bolgna controlled the fruit business in Detroit, and we gave the concession to him. He took LaMare in with him—whether by choice or not, I do not know.

Fred H. Diehl, head of the purchasing department, was so outraged at this arrangement that he resigned from the company. However, there was no more violence over the fruit concession after that—not to this day.

Mr. Ford made me his agent in dealing with the underworld. He gave me a job that a number of times almost cost me my life.

The reason I was able to deal with all these underworld people for so long and still stay alive was that I always kept them obligated to me, but never became obligated to them. If I had ever accepted even the smallest favor from one of them, I would have become their man; and eventually that would have been the end of me. But even with this policy, a number of times we got into ticklish situations that I survived only by great good luck.

At one time some hoodlums got the idea

that I was working with the law against one of their rackets. They apparently decided that the way to stop me was to get rid of me. Toward this end they had been observing my customary route to the plant for a week, though I did not know it at the time.

One morning when I was about to leave for work I had a strong hunch that it would be wise to take a different route than the one I usually traveled. I was on the point of so directing my driver when I suddenly decided I was being foolish, and said nothing.

But my intuition was correct.

As we drove down a continuation of Ford Road at about seventy-five miles an hour, these hoodlums put a car that they had concealed on a side road in gear, and let it run out into the road in front of my car. There was no way my driver could stop. I heard a loud *zing* in my head as we hit the other car and that was the last I knew.

They took me to the Ann Arbor Hospital. There I recovered, though to this day I bear on my face the scars of this attempted murder.

Once I was out of the hospital, these hoodlums called me up and threatened to get me again. "We'll catch up with you yet," they promised.

Later, however, most of the gang went to prison on other charges.

Perhaps I should explain here why, although attempts on my life were made a number of times, I never prosecuted.

For one thing, if we were going to keep in touch with the underworld, we had to do so by acting within their own code. If I had ever begun a legal action against a member of the underworld, from that time on we would never again have been able to get information from any of them.

A second reason for this, and one that was just as strong, was the fact that Mr. Ford never wanted any of us to enter into prosecutions. When this subject came up he often said to me, and to other officials in the company, "We live in a big glass house here—we can't throw rocks."

In fact, not only did Mr. Ford keep us from any prosecutions, but also he never wanted us to make any public explanations of anything we did. Many times I wanted to explain to the press certain things that occurred, but Mr. Ford's contempt for public opinion prevented this. Mr. Ford even got angry when you tried to explain something to him. He'd say, "Aw, what are you trying to alibi?"

Harry Bennett's basement office was the public relations center for the Ford Motor Company. Newspapermen came to Bennett as both Henry and Edsel Ford were unavail-

able. Henry made his "security" officer the point of contact for the public—greatly reinforcing Bennett's position.

Twice more I was nearly killed because of our dealings with the underworld, both times because we intervened in kidnappings that took place in Detroit. Both cases, which happened within a period of a few months, were notorious in their day. But this writing will be the first time the real story of either has ever been told.

Chapter 10

THE FIRST kidnaping case in which I became involved was the abduction of David Cass, twenty-three-year-old son of a wealthy Detroit family.

Young Cass was kidnaped during the summer of 1929 by a gang known as Eagan's Rats, who came out of St. Louis and were led by Killer Burke and Red O'Reardon.

When the gang took Cass, they also took a young man named Fetterley, who was driving for the Cass family. It so happened that Fetterley was the son of a Ford employee. Fetterley's father came to me for help, asking me to try to get his son back. This gave us a temporary interest in the case—temporary, because the kidnapers soon let Fetterley go. However, newspapermen kept coming out to the plant to see if I could give them any leads. I had none, but I guess the repeated visits of the reporters, plus our interest in young Fetterley, excited the suspicions of the gang.

One night shortly after the kidnaping, I was driving home in a Lincoln when I spotted a parked car at a curve in the road. I knew that the police often waited at this point to pick up bootleggers and hijackers, and assumed that these were police. Then I saw their lights dim and go out—and I wasn't at all sure. I slouched down in my seat as my car drew near them.

It was a lucky thing I did, because as I drew up to the parked car a double-barreled blast from a shotgun blew out my windshield. I stopped the car and got out. It was then I saw there were four men in the car. One of these men walked up to me and shoved a shotgun into my belly. He said, "You S.O.B., you're looking for information, aren't you?"

The man on the other end of the gun was Joseph A. "Legs" Laman, who was acting as contact man for the gang.

I talked like sixteen lawyers for what seemed like eternity but could not have been over ten minutes. I told him we had been interested in getting Fetterley back, because he was the son of one of our employees, but that beyond that we had no interest in the case. Those men didn't have long to spend there, because that was a busy highway, and soon it seemed that I convinced Laman, because they drove away, leaving me alive.

The kidnapers had demanded of Gerson C. Cass, father of the abducted young man, a ransom of $25,000. In a few days, however, this demand was reduced to $4,000. It was taken for granted that this meant, in all probability, that young Cass had been killed. Nevertheless, the father paid out the money as directed. But Laman, though already known to the police as contact man, was shot and critically injured when he came to pick up the money.

Laman was convicted of extortion and given a short prison sentence. Late in October the body of young Cass was found on the banks of the Flint River near Lapeer, Michigan. The jaw had been broken, and a table scarf was knotted about his neck.

I will come back to Laman later; for by this time I was already involved in the second kidnaping case.

In September of 1929 a five-year-old named Jackie was kidnaped in Detroit. His distraught father came to Mr. Ford's office and asked his help.

Mr. Ford had a deep love for children—any children. His prejudices stopped short when it came to children; he never hesitated, for example, to have Jewish children in Greenfield Village School.

Time after time, Mr. Ford would read in the papers about a child or perhaps a whole family of children orphaned by a tragedy, and right away he wanted to help. He often had me act for him in these cases; once, when handling three children for him, I found I'd adopted the kids without knowing it.

As a result of his feeling for children, Mr. Ford brought Jackie's father to see me and said, "Don't we know enough of these hoodlums around town to get some help?"

I told him I thought we did, and went to work on the case. Neither Mr. Ford nor I was interested in apprehending the kidnapers or in bringing about any prosecutions. Our sole desire was to get the child back unharmed.

The father, himself, gave us our first lead. He told me he was pretty sure a man named James Fernando had kidnaped Jackie. He didn't tell me his grounds for his belief. But I had reason to feel it was well grounded, in all probability.

I gave that lead to some underworld characters who were in power at that time. Shortly they informed me that if the police were kept out of the case, they could bring the child back—provided there was no ransom money paid out. This last point was just as important with them as the question of keeping the police out. Maybe they felt that if money were paid out, the child, being no longer useful, might lose his life; or maybe they felt that if money were involved, there was danger of being implicated in the whole thing themselves.

At this point, the father received an anonymous note warning him to stay away from the Black Hand Gang. He gave this note to me.

The note convinced us that we were on the right trail. Otherwise, there would have been no point to sending a threatening letter. We showed the note to the police, in order to try to convince them that we were making progress, and to keep them off the case.

Everything was working out fine, and would have come off all right, except for the fact that suddenly the father lost his head. He completely disregarded our orders not to pay out any money. On his own hook, and without consulting us, he got in touch with a man who couldn't help him at all and paid the man $20,000. At the same time, he began working with the police—though all the while assuring us that he wasn't.

The group whom we were working through learned immediately of the payoff. They pulled out of the case.

However, either to show their good faith or to make sure that they would not be implicated in the ransoming of the child, the gang brought to my office the money that the father had so recklessly paid out. They had got all but $100 of it back from the man to whom it had been paid by torturing him. When they did that, of course, they endangered my life, since the man, a professional killer, was sure to hold me responsible for his torture.

I called the child's father to my office. The group I was working with was there, and Sorensen, too.

I said to the father, "I thought I told you not to pay any money."

"I didn't pay out any money," he protested.

I could feel my anger rising. This man had endangered the life of his child, he had endangered my own life, and now he had the effrontery to lie to me. I said, "You paid out twenty thousand dollars, and I've got it right here."

I put the money on my desk.

The man looked at the pile of bills in silence a moment. Then, still without a word, he walked over to my desk, picked the money up, and counted it through twice.

"There's a hundred dollars short!" he yelped.

I lost my temper—as I think anyone else would have done. I leaped out of my chair. I pushed the man and his money out of my office as violently as I could, and told him to stay away from me.

As far as I was concerned, I was through with him.

Meanwhile, however, Fernando, the real kidnaper, had sent a demand for a $5,000 check to be delivered to him as ransom for Jackie. I don't know whether the fellow had a screw loose or what, but he actually asked for a check. At that time Max Wolfoegle and Inspector Norval Marlett headed the Detroit Police Black Hand Squad—a department no longer in existence. They took a check to Fernando and got the child back. Knowing of our work on the case, they brought Jackie to my office that night.

I called up the father and he came to the plant. I turned Jackie over to him.

Fernando was soon arrested, held on bail of $1,000,000, and subsequently sent to prison.

As far as Mr. Ford and I were concerned, we had helped get the child back, and the matter was now closed. This feeling was not shared, however, by the man who had received the $20,000 and then been forced to give it back.

I got three phone calls from this killer, asking me to meet him. I suppose he wanted money. But, understandably enough, I declined his invitations.

A few nights after his last phone call,

this man drove out to my house with a girl. I was sitting in a chair facing the glass-paneled front door, reading a newspaper. I was expecting a visit from Inspector Marlett, and when I heard the door rattle I called to my oldest daughter to answer the door. Then I dropped my paper and looked up—to see the killer drawing a bead on me. Before I could even rise he fired twice through the door—one shot missing, the other shot hitting me in the side.

The man then fled. Perhaps his gun jammed, or perhaps my daughter's screaming frightened him away.

Inspector Marlett appeared shortly afterward. I had scarcely told him about what had happened when some reporters arrived. We told the newspapermen that we had been cleaning a gun and that it had gone off, breaking the glass in the front door. But one of the reporters was smart.

"Don't give us that," he said. "The broken glass is on the *inside* of the door. Those shots were fired from outside." We could scarcely deny it, and so some of the story got into the papers.

I discussed the shooting with Marlett later. My wound was not a serious one, and for reasons already explained, I told Marlett that I didn't want to prosecute.

As it turned out, that wasn't necessary.

The gunman was later shot and killed on a Detroit street—in some gang feud, I suppose.

Now, to get back to Laman again.

In November, Laman was brought up for trial once more, this time in connection with another kidnaping case—the abduction in April of Fred Begeman of Wyandotte, Michigan, a retired bootlegger. Laman was convicted of having participated in this kidnaping, and was sentenced to from thirty to forty years in Michigan State Prison at Jackson.

But this was not the end of it. The police were still determined to apprehend and convict the kidnapers and murderers of David Cass, and in a short time they accomplished this by framing Laman.

When Laman went to prison, apparently Red O'Reardon had promised Laman to take care of his wife and daughter. Now two state policemen got hold of Mrs. Laman and the daughter. They mussed the two of them all up, and then took them to see Laman. They told Laman, "This is how much Red O'Reardon thinks of you."

Believing he had been double-crossed, Laman turned state's evidence and pointed out everyone involved in the kidnaping of Cass. As a result, six men were sent to prison.

While Laman was testifying before the grand jury in the proceedings that led to the conviction of these six men, we observed that he was in considerable pain. I had Frank Holland spirit Laman out of jail and take him to the Henry Ford Hospital. There he was operated on, and a bullet was discovered imbedded in his appendix. Laman recovered quickly, and Holland took him back to jail. Up to this writing, no one ever knew that this operation took place.

There was, however, a breath of suspicion abroad. A judge called Holland before him and said he had heard a rumor that Holland had recently entertained Laman at a local theatre. Fortunately, Holland was able to point out to the judge that the theatre he mentioned had been closed for about fifteen years.

Laman's sentence was drastically reduced in return for having given evidence against the others. After a while both Laman and an associate named De Long, who had also turned state's evidence, were paroled to me.

I got Laman in my office. I told him, "You know, the only reason I got you out is that you could have killed me that night, and you didn't."

Laman looked at me coldly and said, "Oh, no. If I'd had another slug in my gun, you'd have got it."

I believe there was something of a silence after this remark.

Still, I thought that perhaps out of gratitude for his parole, Laman might be useful to us. At that time there were a lot of bombings going on in Detroit. We had been tipped off that there was to be an attempt to blow up a dam of ours. I asked Laman to get me any information he could on this.

But Laman said, "I'll never be a cop—even if they put me back in jail."

Chapter 11

OUR CONTACTS with the underworld often led people to ask us for help. The kidnapings I have told about were the extreme cases. There were other cases that were serious but not that serious, and often, strangely enough, had their moments of comic relief.

For instance, there was the time the governor of Michigan met Leo Cellura at lunch.

In July of 1930 Leonard Cellura, known to the underworld as Black Leo, allegedly shot and killed two Chicago gangsters named Cannon and Collins. He had been tipped off by a Detroit policeman that this pair were coming to kill him.

Leo hid out for six years. Then in July of

1936 he decided to give himself up for trial; but he wanted to walk in, he didn't want to be taken. He called me at the Ford Motor Company to see if I could get the prosecutor to let him walk in. His call to me, incidentally, was unsuspected by his attorney, who was the brother of the police commissioner.

I told Cellura I'd do what I could to help him, and invited him out to the plant to have lunch with me.

When I invited Cellura out, I completely forgot that the governor of Michigan had an appointment to lunch with me and then meet Mr. Ford. Leo and I were already at the table in the executives' dining room when the governor and an aide walked in and approached our table.

So I did the only thing I could do. I calmly introduced them. "Governor," I said, "I'd like you to meet Leo Cellura."

The governor and his aide sat down slowly, like two men being overcome by rigor mortis, their eyes popping out of their heads. They were both in a great state of agitation, and apparently unable to eat. The governor was known as an excessively proper gentleman, and I got a lot of secret amusement out of watching his agitation.

Soon the governor could stand it no longer, and said that he wanted to make a phone call. I knew that what he wanted was to get away from the table and Cellura's company, so I sent him to a small private office. A few seconds later the aide gulped, "I wonder if the governor needs any help," and bolted after him.

I excused myself from the table and followed the two men. The governor, almost apoplectic by now, said, "Do you realize that man is wanted for murder?"

"Oh, sure," I said.

"Well, this is a hell of a trick!" the governor exploded.

I said, "He's giving himself up. And anyway, he's been around here the last couple of years. They could have got him any time they wanted."

Later, Cellura lost his self-defense case through his ignorance of the English language. But I don't suppose I'll ever forget the performance put on by the governor and his aide because of their unexpected lunch partner.

Again, there was the time I received a visit from a man named Niebiola. Niebiola was the owner of one of Detroit's most famous Italian restaurants. He came to my office one day to ask my help. He said that some gangsters had been demanding he pay for "protection," and that he had refused. He said he could "protect" himself, without any help—which was probably the truth, since he was a big, tough guy. However, he told me the gangsters had now threatened to blow up his restaurant at dinnertime on a certain night.

"Well," I said, "what do you want me to do about it?"

"Mr. Bennett," he said, "they would never blow the place up if you were there. I want you to come and have dinner in my restaurant that night."

"Thanks," I said, "but I'd rather eat somewhere else."

"'No, you come," he said. "Then nothing will happen."

He kept after me until finally I agreed.

I went to Niebiola's on the appointed evening. I had just begun my meal when Joe Tocco, a well-known member of the Detroit underworld of that time, and with whom we had contact, came walking in. He looked around the place and soon spotted me. He walked over to my table.

"Mr. Bennett," he said, "this is a terrible restaurant. You come with me and I'll find you some real Italian food."

I said, "I like it fine here, Joe."

"No, it's no good," Joe insisted. "You come with me, I'll give you a good meal."

We argued back and forth for a while, and finally I said, "I won't go, Joe, because you want to blow this place up."

"Me?" Joe said. "I wouldn't do such a thing, Mr. Bennett."

"I know that's what you want to do," I said.

"Well, we give this man protection, and now he don't want to pay for it," Joe said, with the injured air of a businessman who has been done out of an honest fee.

"You haven't given him any protection," I said, and we had a long argument about the matter. Finally Joe promised that if I went out with him he wouldn't blow the place up.

When we got outside, I still had my doubts as to Joe's intentions, and I expressed them.

"I'll show you I really mean it," Joe said, and he walked around the building and removed six sticks of dynamite he had wired against the walls.

On still another occasion I had a chance to see how well Joe could act the part of injured innocence—but not for long.

It was during prohibition, and Inspector Mayo, a Federal Revenue agent, came to see me to ask if I'd put him in touch with Joe Tocco. I got hold of Joe and had him come to my office.

The agent said to Joe, "Joe, we've heard you've got a still out at your place, and we're coming out there to knock it over on Saturday."

"Me? A still?" Joe said. "Why, I haven't got a still. I wouldn't do anything like that."

"Now, Joe," the agent said. "Mind, I'm not saying you've got one out there. I'm just telling you we've heard that, and we're coming out."

"I haven't got a still," Joe repeated.

"Well, that's good," the agent said, "because we're coming out Saturday."

"Why don't you wait until Monday?" Joe said. "My biggest business is over the weekend."

"We're coming out Saturday," the agent said.

On Friday night there was one of the loudest explosions ever heard in Wyandotte. Joe solved the problem by just blowing the whole works up.

Joe came to the end of the line in 1938, when he was shot by an unknown assailant. He was taken to a hospital. The police tried to get him to tell who had shot him, but he wouldn't talk. He kept calling for "the boss."

Characters like Tocco and LaMare always called me "boss," much to my embarrassment. They would call me up from a public telephone, leaving the door open so a whole roomful of people could hear them, and, after loudly asking the Ford switchboard operator for Mr. Bennett, they'd yell at me, "Hello, boss!" Most of them didn't speak good English, and this greeting was roughly equivalent to a typical American "Hello, kid." But it used to embarrass me.

The police sent for me, and I went to see

Joe. I talked to the doctor before I went into Joe's room, and was told Joe didn't have a chance.

I said hello to Joe, and then I begged him to stop calling me "boss." I said, "Joe, you're hurting me when you do that."

"All right, boss," Joe said. "Am I going to die?"

"I'm afraid you are, Joe," I said. "The doctors tell me you're all punctured inside."

"Will you get me a priest?" Joe said.

"I don't know if any priest will come to see you, Joe," I said, "but I'll try."

I went to see a priest I knew, and he agreed to see Joe. I took him to the hospital, and he administered the Church's last rites to Joe Tocco.

Because Mr. Ford had come to dominate my life so completely, to the extent that I literally had no existence outside the Ford Motor Company, late in the 1920's I decided to build a home away from Detroit, feeling that might help me find a life of my own. With this in mind, I built a house just outside Ann Arbor, which came to be known as "the Castle."

The Castle was a three-story brick structure with two towers. Actually, the towers had not been contemplated in the original architectural plan, but Mr. Ford suggested them.

A narrow tunnel was built leading some

distance out from the house. There was a gravel sluiceway there when I built the house, and all that had to be done to create the tunnel was to roof over the old sluiceway. This was, again, one of Mr. Ford's suggestions. I have already explained how he liked to inject himself into other people's building and decorating. But he had a special passion for duplicating in my home things that he had in his. Mr. Ford had elaborate tunnels underneath his own home.

In one tower we put a secret door and a winding staircase, which connected with the tunnel. Mr. Ford thought I should have it as an escape for the children, in case they were endangered. However, the secret exit was never used.

At the end of the tunnel I kept my lion and tiger cages.

For several years I raised lions and tigers at home, and often took one into the plant with me. I have always been fond of cats of any kind, and this became a hobby with me.

Naturally, such a hobby led to several amusing incidents.

One time, I remember, a lion I'd brought into my office gave Rockleman and Lou Colombo a pretty bad scare. They had come in to see me, and hadn't noticed the lion. We were about halfway through our conference when suddenly they both spotted the big cat. Without a word, they simultaneously leaped to the top of my desk. It took a little while to get them down again.

Still another incident involved Elmer Hogan.

Hogan was out at my home one evening. As he was about to go, he remarked that he liked lions and wished he had one.

"Why, I'll give you one," I said, and loaded a lion in the back seat of his car.

Hogan started for home. This particular lion was unusually friendly, and also was inordinately fond of riding in automobiles. The first thing Hogan knew, he felt a big paw resting on each shoulder and the lion's head nestled down on top of Hogan's skull.

Hogan didn't know just how friendly the lion was, or what the animal had in mind. He was afraid to try to look around or push the lion off. He said later that his arms began to shake so, he had some difficulty hanging onto the steering wheel.

Hogan decided the best thing to do was to find a cop, and he began driving around looking for one—the lion still with his feet on Hogan's shoulders, his jaw nestled comfortably on Hogan's head, while he looked out the windshield happily at the passing scene. But it was late at night, and Hogan never saw a policeman. At length he went home and drove straight into his garage.

He slipped out of the car and closed the garage door quickly, and decided that the garage was as good a place to keep the lion as any.

After a while Hogan began to see that the lion wasn't dangerous, and got along with him all right. But then one night it escaped from the garage and disappeared.

For a few nights the lion prowled around the streets of Detroit, knocking over garbage cans whenever he got hungry. He was seen at these activities, but no one could believe they'd seen a lion prowling around Detroit, so it was reported to the papers that a "big yellow dog" was raising hob with the city's sanitation.

As soon as Hogan read this report, he knew it was his lion. He got in his car and cruised around the neighborhood where the "big yellow dog" had been reported, and sure enough, he found the lion. He loaded the cat into his car. The whole thing was getting a little too rough for Hogan, so he drove straight to a Dearborn police station and made the police a present. They kept the lion for some time, and then I was told that the lion had "hanged himself."

Around the time I was building the Castle, a new element came into my work for Mr. Ford.

Mr. Ford decided that it was important to build up better relations between the company and public officials, and asked me to concentrate on building up good will for the Ford Motor Company. Since I was just building my home, and didn't have much of a place to entertain in, I got a series of three yachts for this purpose. The last one was the S-Star, a seventy-five-foot steel cruiser rebuilt for me by Mr. Ford. It was eventually taken by the Navy at the beginning of the war.

I began entertaining politicians and other people, at first on my boats, later in my home. We did this entertaining to try to get them to like us, and to keep us in the know. Soon I had to turn my home into a veritable night club. It became hard work for me, and hard on my family.

Entertaining for Mr. Ford became such a large part of my job that I fell almost completely out of touch with what was going on in the plant the last ten years I was there.

Harry Bennett entertained major celebrities. Charles and Anne Morrow Lindbergh were among his many guests. According to Mrs. Lindbergh, the arrangements were lavish.

I was continually spending my own money on entertaining for the company. I'd tell Mr. Ford this, and he'd say, "Now, you aren't going to lose a cent by this." But then he'd never do anything about it. I think he was pleased to see my money going for this purpose, instead of accumulating as a backlog for myself. It was one more way to keep me dependent upon him.

It got so I had to entertain people who were really Mr. Ford's personal guests. He would invite people to stay with him, with all the generosity and hospitality in the world. Then he'd drop them off at my office, and I'd have to put them up for the night.

As part of this mission of building up good will for the company, I always distributed big tips wherever I went. This had a double virtue, in that it also brought me a lot of information. Doormen, porters, and so on often learned things it was good for us to know.

But big tips didn't always work out perfectly. I remember in particular one time during the depression when I was invited to have dinner at the Athletic Club in Detroit.

As soon as we were seated, I called the waiter over and slipped him a twenty-dollar bill. I said, "Now, be sure to take care of this table."

The waiter, whom we later learned had been hired only for the day, disappeared, and we sat there for fifteen or twenty minutes without getting any service at all. I called the headwaiter over. "Where's our waiter, anyway?" I demanded.

"Well, I don't know what got into that fellow," the headwaiter said, shaking his head. "He just went out into the kitchen, hung up his apron, and quit."

Chapter 12

I KNOW I have been referring rather casually to the various attorneys who were employed by the Ford Motor Company during the years I was there, and perhaps it's time now to try to get their succession straightened out.

As I have already said, Alfred Lucking was general counsel for the Ford Motor Company when I began there; he left in the early 1920's.

Then Clifford Longely became general counsel. After a few years he was set up in an office on his own, on a retainer. At the same time Mr. Ford began to go sour on him—largely, I think, because Longely made suggestions for changes in Mr. Ford's will. He wanted Mr. Ford to turn over his money to Edsel immediately to avoid the inheritance tax.

In 1928 Mr. Ford began to take an interest in Lou Colombo, who was a leading criminal lawyer in Detroit. He got interested in Lou because of the skillful manner in which he handled a couple of cases that were in the public eye.

In one of these cases, and the one that really got Mr. Ford interested in Lou, Lou defended a Dr. Loomis, accused of murder. It was a sensational case, which stayed before the public a considerable time. Mr. Ford watched the thing in the papers, following Lou's defense. When Dr. Loomis was acquitted (he really *was* innocent), Mr. Ford was so impressed with Lou that he told me to hire him.

So Lou Colombo became general counsel of the company in 1928, and Longely became Edsel's personal attorney.

To carry the matter on to its conclusion, Colombo stayed with the company until 1940, when he left. He was succeeded by I. A. Capizzi, who remained as long as I was with the company.

Edsel never wanted either Colombo or Capizzi, on the ground that they were both criminal lawyers, and he felt we needed corporation attorneys.

I have already said that Edsel's first bad feeling toward his father came at the time Mr. Ford got Kanzler out of the company.

On Mr. Ford's part, I believe his first bad feeling toward Edsel developed over the question of the estate tax.

Edsel thought that if his father were to die, most of the Ford fortune would pass to the government. He wanted to follow Longely's advice and have Mr. Ford's will changed, with most of the property put in escrow. Both Longely and Crawford, Edsel's assistant, encouraged Edsel in this.

Mr. Ford became very upset over the whole thing, and, I believe, saw it as a plot to get everything away from him. He became extremely hostile to both Longely and Crawford, and laid the whole thing to them. At one point Mr. Ford was so angry he told me he would be damned if he didn't spend every cent he had before he died.

Mr. Ford gave the proposed will changes, as drawn up by Longely, to Colombo to go over. I don't know whether or not Lou knew it, but Mr. Ford's motive in this was to test Lou.

After he'd gone over the whole thing, Lou indicated disapproval to Mr. Ford. Mr. Ford praised Lou to me after that; he couldn't praise him enough. There wasn't any question about it, Lou was in solid. Mr. Ford told me he wanted to have Lou draw up a document that would establish a board of trustees to control the company and carry out his policies for ten years after his retirement.

But Mr. Ford never actually put his desire into effect until Capizzi came to the company—and that's a story I'll tell in its proper place.

Meanwhile, an incident occurred that was not very important, and yet was a fitting thing to mark the end of the bull-market twenties.

In 1929 Mr. Ford had Rockleman and Liebold sell the D. T. & I. to the Pennsylvania Railroad. The deal included a million dollars in cash. Mr. Ford wanted to send this currency to New York to be deposited in the Corn Exchange Bank, but Liebold objected to the expense of hiring an armored express car.

"Well," I said, when Mr. Ford told me about it, "just put it in a suitcase, and I'll take it down."

So I walked into a Pullman compartment in Detroit with a .45 under my arm and a million dollars in currency in a suitcase. With me was one of Liebold's clerks, who was frightened almost speechless. All the way to New York we had a difficult time. He kept slamming the compartment door shut and locking it, and I kept kicking it open so we could get a breath of air.

At Grand Central Terminal we were met by Joe Palma, and the three of us went to the Corn Exchange Bank. There we opened up the suitcase and deposited the million dollars.

At about this time, still another responsibility was added to my work for Mr. Ford. The last twelve or fifteen years I was with him, I helped Mr. Ford give interviews to the press, a job that previously had been completely in Cameron's hands. The change took place gradually. Soon Mr. Ford would see newspapermen only in my office, and nowhere else.

It wasn't an easy job, because Mr. Ford, when being interviewed, often said anything that came into his mind, and what came into his head quite often would have embarrassed the company considerably were it to be printed. It was my task, when Mr. Ford came out with something that would have harmed him, to interrupt with "What Mr. Ford means—" Then I would rephrase what he had said so that it had enough of the original thought to keep Mr. Ford happy, but was altered enough to keep the company out of trouble.

There was at least one pleasant aspect to the job, which was verifying for reporters some of the famous stories that grew up about Mr. Ford, many of which were apocryphal. For instance, he verified as true the anecdote about Barney Oldfield.

Back in the days when Mr. Ford was

trying to get a start as an automobile manufacturer, it was the custom to build racers and try to break a speed record in order to get public attention. Mr. Ford built the "999," which broke all the speed records of the time and set Mr. Ford on his way. At first Mr. Ford drove the "999" himself, but then Barney Oldfield took over the driving assignment and became the most famous race driver of his day.

Many, many years later, when Mr. Ford was already a great industrialist and a multimillionaire, Barney Oldfield and Mr. Ford got together and began to talk over old times.

"Well, Barney," Mr. Ford reminisced, "I guess you and I made each other."

"Yes," Barney said. "But I did a hell of a lot better job than you did."

Generally, Mr. Ford disliked the press, although he did like Dave Willkie of the AP and Jack Manning, editor of the Detroit *Times.* He once said to me of newspapermen, "Harry, they're all a bunch of skunks, and you know what happens to people who play with skunks." And that attitude didn't make my job any easier.

At this time the Model A was on its way out. Sometime around 1931 Mr. Ford began to develop his V-8 engine and a straight six-cylinder one.

Mr. Ford was as good an engine man as they came. His middle name was Power.

Had he stuck to what he knew, and stayed away from the other things he got into, I have no doubt he would have gone down in history as an even greater man.

While Mr. Ford was developing the V-8 and the six, he was extremely secretive about it. He heard a rumor that Rockleman had discussed the engines with Walter Chrysler, and threatened to fire Rockleman if the rumor were true. He had almost everyone in the plant afraid of his job, and got everyone so tense over the whole thing that he achieved the opposite result from what he wanted: Employees were discussing the engine much more than they would have under more relaxed circumstances.

Mr. Ford had me working my head off to find out if there were any leaks. Things came to such a pass that I actually refused to look at the engines myself. A dozen times he tried to get me to look at them, and I said, "No, I don't want to go in there. I don't want to know anything about the things."

Then one day Alfred Sloan of General Motors and another G.M. executive came out to the plant to pay a visit to Mr. Ford. After a short chat, Mr. Ford led Sloan and his companion out into the plant to the everlasting confusion of every man in the Ford plant, and calmly showed them the two engines.

Why Mr. Ford did things like that I don't know. He just did them.

But meanwhile, more important things were happening. The depression was upon us. It had its effect on the Ford Motor Company, as it did upon every corporation and individual in the country.

In the fall of 1929 the bottom dropped out of the stock market and the United States experienced its most severe financial panic. The bull market had boosted stock quotations up as high as four and five times their par value. The collapse came in late October and early November. In September the quotations on fifty leading stocks averaged 311.90; on November 13 the same stocks averaged 164.43. On one day—October 29, 1929—thirty billion dollars in security values vanished in the stock-exchange panic.

Brokers leaped from windows—there were three such suicides in one day—while businesses all over the nation failed and banks went under. Farms were foreclosed at an unprecedented rate, and apple vendors appeared on city streets. Unemployment, already high, reached astronomical numbers.

In 1932 Franklin Delano Roosevelt was elected to his first term in the Presidency. When he took office, somewhere between 30 and 50 percent of the working population of Detroit were without jobs.

Henry Ford fought the depression. He had already increased the $5-a-day rate to $6 a *day. In December 1929 Ford increased the daily rate to $7. He made off-the-record contributions to public institutions and himself subsidized the Henry Ford Hospital.*

While many things happened in the company in this period, I suppose the event that created more publicity than any other at the Ford plant during the depression was the so-called Hunger March.

It was written about and talked about a great deal when it happened, and still on occasion is written about today. Whenever a periodical wishes to write about the Hunger March, it usually illustrates the story with a photograph of me being lifted from the pavement unconscious.

Until now, however, I have never told my experiences in that notorious event, although, inadvertently, I became one of the main actors in the drama played before the gates of the Rouge plant.

In February of 1932 the Rouge plant was going full blast. If my memory is correct, we were the only automobile manufacturers then producing in Detroit. A good portion of the work we were doing—and this comes into my story later—was for the Soviet Union. We were making tractors for them, and we also had a large delegation of Soviet technicians at the Rouge, who were being schooled in Ford methods of production. They referred to themselves as the "Ford Autostroy." They had their headquarters right over Gate Four, on Miller Road.

Late in February we were tipped off that a crowd of demonstrators meant to march on the Rouge. Nevertheless, we did not anticipate any trouble. There had been demonstrations by the unemployed before, both at our plant and in Detroit, and with a few exceptions there had been little violence. Anyway, we went on the assumption that if there was any trouble, the Dearborn police could handle it.

A crowd of about five thousand formed ranks in Detroit. They were led by a group of open Communists. In my opinion, the majority of them were not Communists, however; if they had been, they could have taken over the city. To my knowledge, there were no Ford employees among them. Mayor Frank Murphy gave the marchers a permit to demonstrate within Detroit, and further supplied them with a motorcycle escort, which conducted them to the city limits.

At the line that divides Detroit and Dearborn, the marchers were met by the Dearborn police. The police warned the marchers that if they crossed into Dearborn, they would be dispersed by force. The marchers held a quick consultation and decided to keep going.

As the marchers came on, the Dearborn

police pumped tear-gas shells into their midst. But the wind changed abruptly and blew the gas back at the police. A free-for-all developed quickly, and in a short time the Dearborn police had the devil beat out of them. The marchers continued on toward the Rouge.

Meanwhile, I knew nothing of what had happened at the city limits. At the time, I was reviewing a movie at the Rouge with former Governor Fred Green and some others. Mr. Ford, Edsel, and Sorensen were having lunch at the Dearborn laboratories. The crowd reached our employment office on Miller Road, which we called Gate Four, but which the press usually called Gate Three.

While I was watching the movie, my clerk came into the projection room. He told me there was a lot of trouble at Gate Four. He said that several Dearborn policemen were in the hospital, and that the Dearborn fire department was out there playing hoses on the mob.

It was a bitter-cold day. I said, "I suppose all the reporters and photographers in Detroit are out there, getting pictures of those people being soaked. That's going to be some more nice publicity for the Ford Motor Company."

I then excused myself to Fred Green. I had no other idea than to go out there and stop the fire department, because I knew very well how the whole thing was going to look in the papers the next day. So little did I guess at what was about to happen to me that I said to Fred Green, "Excuse me—I'll be right back."

I went out and climbed into a company-owned Ford sedan, and my clerk, Beach, got behind the wheel. We drove over to where the firemen were playing their hoses. We found the fire chief and I urged him to stop.

"Why," he said, "they're just a bunch of Communists."

I told him, "You're doing just what they want you to!"

He ordered the water shut off, and then I said to him, "I'll go out there and talk with them."

It seems incredible, now, even to me—but I actually thought I was going to pull off a big thing and stop the whole riot by myself.

I told Beach to drive out to the crowd. He was frightened and refused.

"Then get out," I said, "and I'll drive."

With that, Beach drove out toward the mob and right into the midst of them.

I leaned out the window and called, "Who are the leaders here? I want to talk to them."

A man named Joe York answered, "We're all leaders here."

I then made a wisecrack—which was my

mistake. I pointed up over Gate Four, at the headquarters of the Autostroy, where some Russians were actually watching everything out the windows. Rocks flung by the mob were striking there, and I cracked, "You're stoning your own fellows up there!"

A woman marcher dressed completely in red yelled back at me, "We want Bennett, and he's in that building."

I said, "No, you're wrong. I'm Bennett."

And I got out of the car, actually believing, fantastic as it now seems, that I was going to settle the whole thing right there.

As though at a signal, a dark cloud came flying toward me. It was as though a dense flock of pigeons were dropping toward me out of the sky. But there weren't any birds—the dark cloud was solidly composed of chunks of slag.

I grabbed hold of York, thinking I would protect myself that way. A chunk of slag seared across the skin of my forehead and furrowed my scalp. Blood rushed over my face. I fell to the pavement, York on top of me. I tried to get up, but the weight of York's body held me down. My whole body was by now bruised by the slag rained toward me.

York got to his feet. Just as he did so, Dearborn police opened fire on the mob from in front of the gate, and Detroit police, who had come out to see the show, began firing from the overpass. York and four others were killed, and many were injured.

I was determined to get out of that rain of bullets. I got to my feet. Someone hit me on the back of my neck with a chunk of slag. That was the last I knew for some time.

I lay there on the street, unconscious—but alive. And, looking back, I think my life was saved in a most curious manner. If York or someone else hadn't fallen on me, I would have got up at the moment he did, and I would have been shot. I'm also sure that if I hadn't been hit by that slag and knocked out I would later have been killed by police fire.

The timely appearance of State Police Captain Donald Leonard, I was told, was all that prevented further bloodshed. He drove right into the midst of it all and stopped the firing.

But the riot wasn't over by any means. By now, employees were pouring out of the plant, anxious to get on both sides of the battle.

Captain Leonard went into the employment office and got Governor Brucker on the phone and asked for assistance. I'm told the only comedy in the whole affair was the way Leonard really gave the Governor hell.

As for me, I was taken to the Henry Ford Hospital. There, for two days, the doctors futilely treated me for my scalp wound, which at first was thought to have been caused by a bullet. The treatments had no effect. I remained a little out of my mind, and was seized by periodic and violent muscular spasms.

Finally Dr. Badgely of the University of Michigan Hospital came to see me, and found I had a neck injury.

The newspaper stories that came out in rapid succession got more and more fantastic. At one time the papers had me out there facing the crowd with a gun smoking in each hand; another time they had me wading into the mob hurling tear-gas bombs.

During the summer, Wayne County Prosecutor Harry S. Toy conducted a grand-jury investigation into the Hunger March. While the jury criticized the Dearborn police, it found the Ford Motor Company without blame in the five deaths.

In fact, Henry Ford was admired by a large proportion of the American public. His natural simplicity made one of the most successful industrialists into an individual who was perceived as on the side of the common man. He inspired thousands of letters of sympathy and support, many of a religious bent, when he won the libel case against the Chicago Tribune. *When Ford encountered hard times, necessitating severe cutbacks in 1920, he got offers of loans from countless members of the public. Even when Ford's war production in World War II fell behind, the public opinion polls showed Ford as contributing the most of the big car companies—although the contrary was actually the case.*

The public saw Ford as having a mission and despite his detours and reversals, he symbolized the hardnosed, down-to-earth, practical, and patriotic businessman that suited Americans. His successors kept their independence together with that sense of mission. There is still an admiration for Yankee ingenuity and a willingness to believe that there are better ways to reach one's goals. The Ford Motor Company has prospered. So have Ford graduates at other companies. None of these individuals can believe that any other country can, in the long run, outproduce the United States . . .

Chapter 13

IN 1933 an important change came into my personal life.

I have already explained how, gradually, my every waking moment came to be absorbed by the Ford Motor Company, and how my relationship with Mr. Ford assumed a character that left little room for other relationships. I had married and had children, but having a home meant little to me, and I made a mess of my marital life.

Then I met my present wife. I began to understand what I had been missing, and how much a home could mean to me. She helped me to this realization. When I asked her to marry me, she said, "I'm not going to marry the Ford Motor Company. If we're not going to have any home life, I'm not interested."

I went to Mr. Ford then. I told him I was no longer going to work nights or weekends, and that I was going to begin taking vacations. I told him that I wanted to make a home, and if that was impossible working for him, then I would go somewhere else.

Mr. Ford heartily approved my marriage, encouraged me, and agreed to the change in program I had laid down.

However, looking back now, I must admit that my success in declaring my independence was a limited one. True, I no longer worked on the weekend or at night, and I took occasional vacations. But every Sunday Mr. Ford called me at home, sometimes as often as two or three times. And, as I've said, he called me at nine-thirty at night for twenty years, wherever I might be. My wife and I did manage to establish the home life we both wanted. But the fact that my home was soon turned into a night club because of the entertaining I did for Mr. Ford handicapped our efforts, and made what we wanted to achieve a difficult task.

But let's get back again to the times we were living in.

The National Recovery Act was passed in 1933, and for about two years its famous symbol, the Blue Eagle, flew over the land. Among the automobile manufacturers, Mr. Ford was the sole hold-out. He refused to go along.

The National Recovery Act was a try at a planned industrial society, in which industry was supposed to do its own planning and be self-regulated. During the summer and fall of 1933 countless industrialists and executives of all kinds came to Washington and drew up a total of about 750 "codes of fair practice." These codes were supposed to put a voluntary floor under wages, establish a ceiling on hours of labor, regulate prices, and establish "quotas" of manufactured goods for industry.

The program was under the supervision of General Hugh Johnson, known to the press as "Iron Pants" Johnson.

While the Automobile Code was being drawn up, Mr. Ford was invited to a party in New York by Pierre du Pont. I guess the idea was to persuade Mr. Ford to go along with the others.

But the notion of having Pierre du Pont appeal to Mr. Ford was the worst possible idea anyone could have devised. Mr. Ford hated the Du Ponts. He had a kind of persecution complex where they were concerned. He seemed to feel they were out to "get" him. He always claimed they ran our government. He considered this point was proved when one of the Roosevelts married a Du Pont, and he said to me delightedly, "Now will you believe I know what I'm talking about?"

In any event, as Mr. Ford reported it to me later, here is what happened at the party in New York:

"Mr. Ford," Pierre du Pont said, "I want you to go along with us on the NRA. That Blue Eagle is my baby."

"That's all the more reason I don't want to have anything to do with it," Mr. Ford said. And he got up and walked out on the party.

General Johnson couldn't believe that Mr. Ford meant this, and a while later he came out to Dearborn to see Mr. Ford himself. I was invited to sit in on the conversation.

Johnson tried in vain to persuade Mr. Ford to go along. But he told Johnson just what he'd told Du Pont, and added, "I wouldn't like it even if it was good."

Johnson went away still thinking Mr. Ford might change his mind. But he never did. I couldn't see much sense in Mr. Ford's stand; it seemed to me that the Automobile Code gave the manufacturer every break in the world. I guess it was his hatred of the Du Ponts that made him hold out as long as the NRA lasted.

Meanwhile, the financial crisis that had begun back in 1929 was getting worse all the time. It came to a head in 1933. One effect it had within the Ford Motor Company was to bring to a close Ernest Liebold's long reign of power. He remained for some years, but his wings were clipped.

In the first week of October 1931, 166 banks closed their doors. Earlier, on December 11, 1930, the misnamed Bank of the United States, with sixty-two offices in New York City, went under with 400,000 depositors and $203,000,000 in deposits. The banks in Detroit were under severe strain by late 1932, and street runs were imminent. On February 14, 1933, Governor Comstock invoked the Michigan bank holiday, closing all the banks in the state to conserve their remaining assets. The movement spread, and on March 5 President Roosevelt found it necessary to declare a national bank holiday. Every bank in the United States was closed for nine days.

Before I can tell the story of what happened to Liebold, I must fill in a little background. I must explain that before he became Mr. Ford's business secretary, Liebold was cashier of the Dearborn Bank. When he began working for Mr. Ford, Liebold did not drop this connection, and he remained an official of the bank. Throughout the years he worked for Mr. Ford, he remained active in banking and involved in the city's financial circles.

Shortly before the bank holiday was declared in Michigan, I got a tip from a newspaperman on the Detroit *Times* that the Union Guardian Trust Company of Detroit was in bad financial straits. Since Mr. Ford had many millions of dollars there, I told Mr. Ford about my tip at once.

The Union Guardian Trust Company was an operating branch of a financial holding company known as the Guardian Detroit Union Group. Kanzler was chairman of the board of the holding company, Edsel was the largest stockholder in the holding company and also a director, and in addition he was a director of the Guardian Trust Company. Longely, Edsel's attorney, was president of this bank.

Mr. Ford at once asked for a meeting with Longely and Liebold. They met with Mr. Ford, and brought along with them Rudy Reichhardt, a state bank examiner. Reichhardt and Liebold were very friendly.

I don't know the details of this meeting. It was Mr. Ford's purpose to find out by indirection just how stable the Guardian Trust Company was. He didn't trust these men to give him direct information if he asked for it, so he told them that he intended to draw some millions of dollars out of the First National Bank of Detroit, where he also had money on deposit. These men told him that he could never get his money. Then Mr. Ford knew the worst.

When he came out of this meeting, Mr. Ford told me that things looked bad. I had about sixteen thousand dollars on deposit in Liebold's bank, and Mr. Ford suggested that I draw out my money at once. I said it sounded like a good idea, but added, "What will I do with it?"

"Just give it to me," Mr. Ford said. "I'll keep it for you in the kitty."

I should explain here that Mr. Ford had a safe at the Dearborn laboratories, which he called our "kitty." He and I both put there money that we got from various sources; we had in there, at one time, about four million dollars in cash. I think that in other corporations this might be called a "contingency fund," from which expenditures could be made without explanation. There was an understanding between Mr. Ford and me that when he died, whatever was left in the kitty was to be mine. About two million dollars of that money was spent. I don't know entirely how Mr. Ford used it.

I drew my money out of Liebold's bank and gave it in currency to Mr. Ford.

Meanwhile, I tipped off all my friends who had money in Liebold's bank. Liebold always accused me of starting a run on his bank. Well, I don't know; maybe I did. I certainly was more concerned for my friends than for Liebold.

After this meeting between Mr. Ford, Liebold, Longely, and Reichhardt, Liebold disappeared.

There was a great fuss about it in the papers. Liebold hadn't even informed his wife he was going, and she feared he had been kidnaped.

Actually, no such thing had happened.

Where had Liebold gone? He had driven to the northern part of the state. Liebold went first to Carl Schmidt, an old friend of his and a banker at Harris, Michigan. Apparently Liebold hoped that Schmidt would do some favor for him. But it seems Schmidt refused.

After that, Liebold just wandered about the state. Finally he ended up in a hotel in Traverse City, Michigan, where he registered under an assumed name. He went to the bank in Traverse City and got a safety deposit box.

Meanwhile, I was kept informed of every move Liebold made by Donald Leonard, a very capable police officer who was then captain of state police, and now is commissioner.

Leonard talked to Liebold in his hotel room, and Liebold told him that he had just gone away "for a rest." That is what he had told the reporters, too. But Leonard had found out about the safety deposit box. He called me up and asked if I wanted the box opened.

At this point Edsel intervened. He informed me that, before taking flight, Liebold had called him at the Dearborn Inn and had asked Edsel if he wanted Liebold to "put away some money" for him. Edsel had said no. But now, Edsel felt there'd been enough bad publicity. So I told Leonard to let Liebold go, and Liebold came back.

That marked the end of Liebold's reign of power in the Ford Motor Company. After that, Mr. Ford wanted to get him out, and he thought that derogatory newspaper information would make Liebold resign.

Again Edsel intervened. Mr. Ford called me. He said that Edsel was in his office, and thought we ought to stop giving the papers information on Liebold. I don't know what Edsel's interest was, since he never liked Liebold very much. In any event, I answered, "All right, it doesn't make any difference. They've got it all now, anyway."

However, this drastic treatment appeared to have no effect on Liebold. He had no intention of resigning.

When I told Mr. Ford that Liebold wouldn't resign, Mr. Ford said, "All right, we'll just let him sit there and look out the window."

As Liebold dropped out of Mr. Ford's favor, Frank Campsall took over his work. Campsall had been working as a secretary for Mr. Ford for some time. He had never got along well with Liebold, and once tried to resign because of the clashes between them.

But now Campsall took over as Mr. Ford's personal secretary, although he never assumed the power Liebold had for so long wielded. Campsall was always friendly to me, and I had a much better relationship with him than I ever had with Liebold. Campsall remained in this job until 1946, when he died at Mr. Ford's Georgia estate.

As for Liebold, for a while we let him answer letters that were written to the company. Then I discovered that Liebold was sending out replies that were highly inappropriate.

I told Mr. Ford about this. We went to Liebold's office and looked into his files, and Mr. Ford was furious to see what kind of stuff Liebold had been sending out. He instructed me to see to it that Liebold wrote no more letters.

I tried to warn Mr. Ford, as I had many times previously, that he ought to go all the way and get Liebold out of there. I told him: "Liebold will turn on you."

"Oh, no, he won't," Mr. Ford insisted.

But he did. When I told Liebold that Mr. Ford had ordered him to stop answering letters, Liebold got angry and made a violent statement about Mr. Ford.

I reported this statement to Mr. Ford before several witnesses. "Why," he exclaimed, "the dirty ———"

I was delighted at this reaction, which had been my own all along, and I said, "Do you want me to go back there and throw him out?"

"Naw," Mr. Ford said, his manner changing immediately. "You just cool down, and I'll talk about this with you a week from now."

That was a way he had of getting out of things with me: to pretend that I was acting solely out of temper and impulsiveness. But I observed that when a week had passed, he never brought up that particular subject.

Later, Mr. Ford told me that he didn't want to throw Liebold out because if he did he'd be pleasing too many people he didn't like. That was the motive he gave; but it wasn't the real one.

The plain truth of the matter is that Mr. Ford was afraid of Liebold.

Chapter 14

PERHAPS this is as good a time as any to interrupt the march of events that took place within the Ford Motor Company to throw what light I can on Mr. Ford's home life. I never had much to do with Mrs. Ford and never knew her especially well, so I must admit to being no expert on the Fords' marital happiness. However, inevitably certain things did come to my attention.

Clara Ford was, to use a delicate word, a very frugal woman. She used to darn Mr. Ford's socks. I suppose there wasn't anything wrong with that, but as it happened, Mr. Ford detested nothing in the world so much as a darned sock. He claimed they hurt his feet.

Many times when he was with me, Mr. Ford would have me stop the car in front of some store and ask me to go in and buy him a pair of new socks. Then he would change in the car, tossing the pair Mrs. Ford had so carefully darned out the window. The spectacle of a man with a billion dollars changing socks in his car so that his wife wouldn't know about it was, at least, a unique one.

A second evidence of Mrs. Ford's frugality arose one time when I pointed out to Mr. Ford that he had done many kind things for people that no one knew about. "Why don't we publicize that sort of thing?"

Mr. Ford said, "Well, if we do, there'll be more after us." He added, "And anyway, I don't want trouble at home. You know how my wife is about money."

Once, at least, I was called upon to solve a little domestic trouble Mr. Ford had got into, and my memory of that one is rather vivid, since I failed in my assignment.

There was a Finnish servant girl working at the Residence with the remarkable name of Wantatja. One day this girl went walking in the garden, presumably while resting between tasks. There, quite by chance, she saw Mr. Ford. He was standing behind a hedge, patting the hand of a second girl servant named Agnes, who was crying bitterly. Mr. Ford happened to look up from his consolations and saw Wantatja.

Mr. Ford told me about the whole thing. He said he had only been comforting Agnes because of some grief. Nevertheless, he said, he was afraid that Wantatja might get to Mrs. Ford and recount what she had seen. Therefore, Mr. Ford suggested it might be wise if I were to send Wantatja back from whence she came.

I managed to ship Wantatja back home without any fuss. But then her brother came down to Dearborn and tried to see Mr. Ford. I didn't let him get to Mr. Ford, and before he was through he went to both the FBI and the Michigan State Police, who, if they wish, can verify this story. Finally I got rid of the fellow.

All this didn't do much to quiet Mr. Ford's nerves, and he now thought it would be a good thing if I could get rid of Agnes, too.

I learned that Agnes had a brother of whom she was fond, and who was anxious to settle on the West Coast. That simplified the whole problem. I got Agnes's brother a good job on the West Coast, and then, without much difficulty, talked Agnes into going out there with him.

I now thought the whole matter was ended, and that I had successfully helped Mr. Ford out of his jam. But such was not the case.

Some while later, Mr. Ford called me

and asked, "Is Agnes happy out on the West Coast?"

"Oh, sure, she's all set out there," I said. "You'll never hear any more from her."

"Oh, won't I?" Mr. Ford said. "Well, she's right here in Dearborn."

He then explained that Mrs. Ford had heard whisperings from other servants that had excited her suspicions. She had hired a Pinkerton detective, who traced Agnes and brought her back to Dearborn.

Mr. Ford concluded, "Well, this time I'll handle it."

And I guess he did, because I never heard any more about the whole thing.

Other incidents in Mr. Ford's family life into which I was drawn inadvertently were just sheer comedy.

Mr. Ford had a long running feud with a certain woman. She was very close to Mrs. Ford, who liked her a great deal. I think Mr. Ford was jealous of Mrs. Ford's relationship with this woman.

At one time this woman wanted to see Mr. Ford about something, and kept coming to his office. But Mr. Ford didn't care what she wanted. He just didn't want to see her. Every time she came to his office, he slipped away.

This game of tag went on for some time, while, I imagine, the woman's exasperation mounted.

Whether what followed was connected

with the above or not, I don't know. But here's what happened:

After Mr. Ford had successfully eluded this woman for several weeks, I one day got a phone call from Frank Campsall. He asked me to come over to Mr. Ford's office, because he had received a suspicious-looking parcel in the mail.

I went over to Dearborn. There I found Mr. Ford, Campsall, and a few others, standing around a table regarding morosely a beautifully wrapped parcel. Their attitudes indicated they considered it to be a bomb.

I didn't think there was much likelihood of that, and while the others backed up a few steps I opened the parcel. There was a handsome box inside the wrapping, and in the box a neat pile of manure. There was also a card.

I burst out laughing after reading the card, and I handed it to the others. They all read it, but no one even smiled. They just stood there, looking at each other.

I was disgusted, and I said, "If you haven't enough sense of humor to see that's funny, the devil with all of you." And I walked out and went back to my office.

Well, Mr. Ford learned that this woman was a devotee of spiritualism, and that she put especially deep faith in a certain medium in Detroit. So he worked out a plot.

First we got something on the medium, and then we hired her. We instructed her to tell this woman that at ten o'clock on a certain Tuesday night she should go out on her bedroom balcony; there she should raise her arms three times to the moon and make a wish. If she did that, the medium promised, the wish would come true.

At a little before ten on the appointed evening, Mr. Ford got Mrs. Ford into the car by saying that he felt like going for a drive. Probably to Mrs. Ford's wonderment, Mr. Ford parked the car under the woman's balcony, saying, "It's a nice night—let's just sit here and look around."

A few minutes later, promptly at ten, the woman appeared on her bedroom balcony. She slowly and gravely raised her arms to the moon three times, and silently made her wish—probably that Mr. Ford would drop dead.

Mrs. Ford watched the performance with astonishment, and Mr. Ford craned his neck and looked surprised.

"Well," Mr. Ford said, when the woman had returned to her bedroom, "she certainly takes that stuff seriously, doesn't she?" And he drove off.

It is well known how Mr. Ford became preoccupied with the past, and built up Greenfield Village, in Dearborn. I had nothing to do with this, and so have nothing to add to already published accounts. For many years the overall management of the Ford farms, which included Greenfield Village, was in the hands of Ray Dahlinger. Mr. Ford admired Mrs. Dahlinger a great deal. Mrs. Ford did not.

Mrs. Dahlinger was Mr. Ford's secretary for some time, and it was while she was holding that position that Dahlinger met and married her. Mr. Ford did a great deal for them, including the building of an expensive house.

However, once the house was built, the Dahlingers wouldn't move into it. Ray Dahlinger claimed he was reluctant to move in because, if they did, her folks would move in with them. Mrs. Dahlinger, on the other hand, maintained the real trouble was that if they moved into the house, *his* mother would move in.

Mr. Ford got very annoyed with the whole situation, and told me that the Dahlingers either had to move into that house or get out. To settle the matter, I built a home for Ray Dahlinger's mother, and then the Dahlingers moved into their new home.

There was always something in my make-up—I don't know what it was—that attracted me to the great, and especially the great athlete. As well as I can put it, I was attracted to the man who could do what I couldn't.

Once I got to know the great, of course, they turned out to be just human beings like anyone else. But my admiration and sympathy remained.

It's an old story with those who reach the pinnacles. They get up there, and for a while they are on top of the world. Then, suddenly, someone comes along and knocks them off. The world quickly forgets its admiration for its past hero; everyone's eyes are on the new man who's up there, and the man who was up there before him is simply forgotten, if he's lucky, or is scorned and reviled if he's not.

I know that sequence of events is as old as the world, and will probably go on. But no matter how often I saw it happen, it always moved me. When I saw a big man knocked down, I had a strong impulse to help him.

All of which is a necessary prelude to a discussion of my friendship with Harry Kipke.

It was because of my admiration for athletes that I became a close friend of Harry Kipke, football coach at the University of Michigan. During the early and middle 1930's I tried to help Kipke build better football teams. He told me that he was having trouble building first-rate teams because of a lack of good material, and I did what I could to help him attract more promising men.

What I did to help Kipke was to give summer jobs to boys he brought to me. We'd hire them as gate guards or something like that, and then we'd give them an hour off each day to practice in Dearborn.

Kipke turned out some great teams. But at the end of the 1937 season, Kipke was dropped as head coach by the U. of M.'s Athletic Board of Control. A whole chain of events led up to this action.

For one thing, during the preceding few years the coaches at the university had been getting salary raises and the teachers had not. This made the professors jealous and bitter against Kipke. Also, they felt that under his regime too much of the university's attention was going into athletics, and too little into education. Opposition to Kipke also arose among his own coaches.

To try to decide, at this late date, whether or not Kipke's firing was merited would be beating a dead horse. Whatever the merits of the case, Mr. Ford and I liked Kipke, and decided to help him.

We built Kipke a house near Ann Arbor. We gave him two boats, a sailboat and a cruiser, one of which he later paid for. I also gave him two accounts to handle for the Ford Motor Company—the Eagle Ottawa Leather account and a lunch account. After that, he got the Sparks Worthington account for himself.

Probably this business of accounts is peculiar to the automobile industry, and needs a little explanation.

Suppose that the Ford Motor Company is buying its leather for upholstery from the Eagle Ottawa Leather Company—as we were. We wanted to give Kipke this account, so we simply notified the Eagle Ottawa Leather Company that henceforth we were buying our leather supplies through Harry Kipke. The Eagle Ottawa Company then appointed Kipke their sales agent in dealing with the Ford Motor Company, and Kipke acted as their representative with us and got a commission on all sales.

The two accounts we gave Kipke, and the one he got himself, proved much more lucrative than I had anticipated—and more than Kipke admitted. But after a short while Kipke boastfully showed his bank-book to a Detroit advertising man. He had $100,000 on deposit.

When I learned this, I felt Kipke was expecting too much. I suggested that he give up his lunch account. He did so. Later I asked Kipke to drop a second account, and again he did as I asked.

But Kipke still did all right. Meanwhile, we were frequent social companions. When my wife and I went on a Caribbean cruise, Kipke and his wife went along.

In 1939, at my suggestion and at my expense, Kipke ran for a place on the University Board of Regents. There was considerable opposition to Kipke's candidacy, but thanks to the help of Frank McKay, Republican boss of Michigan, Kipke was elected, to the chagrin of all his enemies within that seat of learning.

Altogether, it was a warm, friendly relationship between Kipke and myself.

Chapter 15

I SUPPOSE that more has been written on Mr. Ford's relations with organized labor than on any other aspect of his career. And I believe I have reached the point in my story where it would be appropriate to go into that. For it was in 1935 that the CIO was organized, and in 1937 the first big strike occurred in the automobile industry.

Up until 1935, American trade unionism was dominated by the AFL, whose interest lay almost exclusively in the craft or skilled worker. In that year, led by John L. Lewis, president of the United Mine Workers, the Committee for Industrial Organizations was set up within the AFL, with the avowed objective of organizing the unorganized mass-

production industries into industrial, rather than craft, unions.

Friction quickly developed between the leadership of the AFL and the CIO, and in 1937 the AFL instructed its state federations to purge themselves of all CIO unions. As a result, in October 1938, the temporary committee was transformed into the Congress of Industrial Organizations. John L. Lewis remained president of the CIO until 1940, at which time he resigned as the result of having unsuccessfully backed the Presidential candidacy of Wendell Willkie. Philip Murray then became president of the CIO.

Meanwhile, the United Automobile Workers had been formed within the CIO. In January and February of 1937 the UAW conducted a sit-down strike against General Motors, which ended February 11 with a contract signed by the two parties. It was the UAW's first big contract in the automotive field.

The UAW got into the automobile industry for some very good reasons.

It was the custom in the industry, when changing models, to pay everyone off and close down the plant. Upon reopening, all the men would be rehired at the flat rate. Thus, a man who might have worked his way up to $8.00 or even $10.00 a day, would be rehired at the wage of a beginner.

This custom, practiced by the Ford Motor Company as well as by all other automobile manufacturers, was the real motive power for the union movement. If I had been one of the men in the shop, and this had been done to me, I'd have been in sympathy with the union myself.

Furthermore, unions were necessary in the automobile industry in order to stop all the politics, the favoritism, and the discrimination that went on in the plants.

However, having said this much, I will say frankly that we didn't want the CIO in the plant. In this, I think, we were not distinctive. Our feelings were not any different from those of the rest of the automobile industry, or, for that matter, of industry as a whole.

Homer Martin was the first president of the UAW. Homer had once been a Baptist preacher, and then, during the depression, had given up the ministry and gone to work in an automobile assembly plant in Kansas City. Homer was a great orator, and did his best job on the radio.

By 1938, when G.M. and Chrysler had already been signed up, Homer turned his radio orations on the Ford Motor Company. He gave the public a grossly exaggerated picture of working conditions at the Ford Motor Company, and complained that I refused to see him.

Finally this stuff got under my skin. I took three husky men and went up to union headquarters to see Martin. When we walked in, I said, "Come on, we're going for a ride."

"Oh, no," Homer said. "I don't intend to be kidnaped."

"All we want," I said, "is to have you come out to the plant and see conditions for yourself. You've been saying you couldn't get to see me. Here's your chance."

"All right," Homer said, "but I'll come out by myself—if I can get in."

I assured him that he could.

A few days later Homer came out to the Ford plant, bringing with him a bodyguard. Mr. Ford came over to meet Homer in my office, and talked with him at some length.

I took Homer to lunch. Then I let him make a tour of the plant. Men could hardly believe it when they saw him walk down the line, and his appearance there caused such a sensation that it almost stopped production.

Homer saw that conditions at the plant were not the way he had believed, but he remained suspicious and unconvinced, feeling that we had put on a show for him. It took some time to get Homer to see things the way they were.

Once Homer got the idea that it was possible to get along with us, he went to Washington to sell John L. Lewis on "peace with Ford." Homer later reported to me that he got a cold reception. Lewis told him, Homer said, to go to Kansas City and make some more radio speeches, adding, "That's where you belong."

Meanwhile, the UAW was by this time raging with factional disputes. This intra-union warfare was at least in part stimulated by a man named John Gillespie.

John Gillespie was an indescribable man. He was known in Detroit for many years as a politician. I want now to make clear his relationship with the Ford Motor Company and with me, because this has been distorted in the press so many times over such a long period. The newspapers have often referred to Gillespie as "Bennett's friend," but that is just exactly the opposite of the truth. Actually, Gillespie and I became bitter enemies.

Up to this point in my story, I scarcely knew Gillespie. A former politician, he had previously done some work for Mr. Ford, particularly acting as bondsman on the tunnel from the Detroit River to the intake at the Rouge. But this had been without my knowledge.

Now John Gillespie disappeared from his home in Detroit. There was quite a lot about it in the papers, because he disappeared in a spectacular manner. He went up to Clare County in northern Michigan and hid—evidently profoundly fearful of someone. He seemed to be in a disturbed condition. "Someone" was after him.

In a short time this persecution complex fastened on Frank D. McKay, Michigan's Republican political boss, and Gillespie laid all his trouble to McKay. Gillespie left his hideout and went to Grand Rapids, where he began raising the devil with McKay. He threw stench bombs into the homes of McKay and Governor Fitzgerald, and began threatening McKay.

A Detroit newspaperman who was in Grand Rapids called me up from state police headquarters there. The reporter informed me that Gillespie was being held by the police, and asked me to help John. He said that the state police commissioner was beside him at the phone, and if I'd tell the commissioner that I'd place Gillespie in the Henry Ford Hospital, the commissioner would put John in the reporter's custody.

When this call came in, Mr. Ford happened to be in my office—which was not much of a coincidence, since Mr. Ford was there much of the time. I turned to Mr. Ford and repeated the reporter's story. Mr. Ford said to me, "You do anything you can for John. And I want to see him when he gets here."

So I spoke to the commissioner and arranged for Gillespie's release.

I did not know it at the time, but later I learned that Mr. Ford felt he was eternally indebted to John Gillespie. The reason for this went back into Mr. Ford's early years. When Mr. Ford was still in the Highland Park plant, Gillespie was water commissioner of Detroit. To help Mr. Ford, Gillespie ran a water main from Detroit out to Highland Park.

I have heard Mr. Ford say to Edsel, who had no use for Gillespie, "We wouldn't be in business if it weren't for John."

Whether Mr. Ford's attachment to Gillespie was all pure gratitude, however, I can't say. Gillespie always let you know what he knew.

When Gillespie arrived at my office with the reporter, he had a week's growth of beard, he needed a haircut, and his clothes were disheveled. I told him to go out and get cleaned up and then come back, because Mr. Ford wanted to see him.

In a few hours Gillespie returned looking a little better. Both Mr. Ford and Edsel came down to see him. They shook hands, sat down, and talked with him over an hour.

Mr. Ford asked Gillespie, "Why do you want to go to the hospital?"

"I don't want to," he answered. "They think I'm crazy."

"Well, you don't have to worry about anything," Mr. Ford said. "We'll take care of you." And then he told Gillespie that he was starting on the payroll immediately, and that there would be things for him to do.

It was pretty hard for me to control Gillespie after that, even though I knew that he could bring us trouble.

The reader will understand that there were few things I wanted so much as to get rid of Gillespie. Finally, I seemed to have the opportunity to do that. A Detroit reporter called me and told me some things about Gillespie I hadn't known before.

With that information, I succeeded in getting the man off the payroll.

However, Mr. Ford promptly intervened. He said to me, "The trouble with John is that you're not paying him enough money. Why don't you give him the insurance here?"

Evidently Gillespie had already talked to Mr. Ford about this, because Mr. Ford went on to say that Gillespie could save us $80,000 a year on our Dallas insurance.

I gave Gillespie our Dallas insurance, since I had no other choice, and I must admit he made good on his promise. After that, he was given all our insurance to handle.

Of course, all this meant that my troubles with Gillespie, rather than being ended, as I had hoped, had merely begun.

I tried for a while to keep from letting him see me, but even that didn't work. Gillespie would come to my outer office and ask to see me. When he was informed that I was busy and couldn't see him, Gillespie would promptly go over to Dearborn and get in to see Frank Campsall. He would say, "I don't know what's the matter with Harry. I want to see him about something for his own good, and he won't see me."

Campsall would get this to Mr. Ford, and soon Mr. Ford would be on my phone.

"Now, Harry, why don't you see John?" Mr. Ford would say. "He only wants to help you."

Gillespie anticipated this call from Mr. Ford to me, and by the time Mr. Ford telephoned, Gillespie would be sitting in my outer office again, waiting for the call to come through.

Resigned, I'd hang up the phone, open my office door, and call Gillespie in. It invariably turned out that what he had in mind was a scheme of some sort that would make money for John Gillespie.

Probably the best way to show just how much of a headache Gillespie was to me is to get a bit ahead of my story and tell about John and his secret cache of tear gas.

Sometime in 1941, not long before our strike, I was sent by the accounting department a $1,000 voucher to sign. It was an expense account for John Gillespie.

I went to the office of Craig, the company treasurer. I told Craig's clerk that before I'd sign the voucher I'd have to know what the money had been spent for. The clerk said the matter was confidential, but he understood that Gillespie had bought some tear gas.

That information, of course, had never been intended for my ears.

I didn't want that stuff around the plant. I considered that it wasn't any good. If you had some trouble and used it, the wind might shift, as it did on the Dearborn police during the Hunger March. And as far as professional criminals are concerned, they just grease up their eyes with Vaseline and the stuff never affects them. And in any event, Gillespie was the last man you'd want to have that stuff in hand.

I promptly got hold of Gillespie and told him I had information that he had bought some tear gas. He virtuously denied it.

"Why, I've only got one eye myself," he said with a mixture of indignation and pathos. "Why would I want to do anything like that?"

But I persisted. "What was the thousand dollars for?"

"That," he answered smugly, "was for a confidential investigation I made for Mr. Ford."

He knew that it was hopeless for me to go to Mr. Ford, because I'd never get an answer. I saw that it was up to me to locate that stuff myself.

I now gathered together about a dozen men who knew the Rouge inside out. We constituted ourselves a searching party, and looked everywhere for that tear gas—every nook and cranny, every tunnel and recess in the plant. We even went over to Dearborn and ransacked Gillespie's office. But we couldn't find it.

Since I couldn't find the tear gas, and since Gillespie obviously wasn't going to tell me where it was hidden, I felt compelled to lay a plot to force the information into the open.

I got hold of a Detroit *Times* reporter and took him into my confidence. I asked the reporter to tell Gillespie, as though he were giving him a tip, that the union was going to march on Gate Ten on a certain day and at a certain hour. I warned him that Gillespie would check on the authenticity of the thing, and told him to have a couple of news photographers at Gate Ten shortly before the thing was supposed to come off, to make it look good.

The reporter agreed to help me, and did as I asked.

At the appointed hour for this mythical "march" on the plant, a little company photographer came dashing into my outer office, breathless.

We had had a film explosion some years

earlier, and had built a concrete vault for storing film right near my office; the keys to it were kept in my outer office.

This little photographer, bug-eyed with excitement, yelled, "Quick! Gillespie wants the keys to the film vault. They're marching on Gate Ten!"

"Whoa," I said. "What's that got to do with the film vault?"

"Oh," the photographer said, "the tear gas is in there."

So I went to the film vault—only a few yards away—and there found about $5,000 worth of tear gas and tear-gas guns. We destroyed every bit of it, some of us using it for sport.

Well, Gillespie, once in the company, began to inject himself into the union situation, working directly under Mr. Ford, and usually at cross purposes with me.

For example, Gillespie began trying to break up union meetings by such stunts as putting tear gas in the radiators of meeting halls. And when John succeeded in something like this, Mr. Ford would rib me about it. "Well, I see John broke up another meeting," he'd say. But I had no part in that sort of thing, which seemed childish to me.

Also, while I was carrying on discussions with Homer Martin, Gillespie injected himself into that situation too, carrying on negotiations with Homer without my knowledge. All that Gillespie accomplished was a lot of confusion. But that was no accident.

Gillespie said to Mr. Ford many times, "The only way to make a deal is to get everything confused. Whenever you can get them all fighting against each other, you can make any kind of deal you want." I guess Gillespie had developed this "philosophy" while in politics, and he carried it into the union dealings he conducted. In fact, in almost any situation, whenever he wanted anything he'd start a fight, and then sneak in and steal the plums.

Many times Gillespie went plunging into a situation saying he represented Mr. Ford, when, to the best of my knowledge, Mr. Ford knew nothing about it. After a while, however, it got so that even when Mr. Ford *did* authorize Gillespie to act for him, I'd refuse to recognize the authorization, and I'd deny to the union men that Gillespie had any such authority. Finally I managed to work things around to where the union leaders wouldn't let Gillespie in, or deal with him if he did get in.

Soon the factional wars being waged within the UAW reached a climax, in which Homer's desire to get along with us was used against him. He was expelled from the union.

After that happened, I got Bill Green and his attorney in my office. I told them that I'd sign an AFL contract covering the whole plant, providing they'd take Homer over. Green agreed to this, and Homer went to work trying to organize an automobile worker's union within the AFL.

Homer worked on this about a year. His efforts were not particularly successful, and finally Green called me up and backed out on the whole thing.

When this happened, Homer was pretty hard up. Mr. Ford came to see me about Homer's plight, and said, "Harry, I guess this is our fault. Let's help Homer."

So we gave Homer a couple of accounts, one with a chemical company. We also gave him a completely furnished home in Detroit. Later Homer traded this home for a farm near Ann Arbor, where he still lives.

After 1933 this business of giving people homes became quite a thing with us. We built over sixty houses for people, after the Wagner Act was passed.

I remember this date particularly because of something that happened then. When the Wagner Act was passed, a reporter came out to the plant and said that he wanted to get Mr. Ford's reaction to its passage. While the bill was up before Congress, Mr. Ford had given out a lot of violent statements on the subject, and I guess the reporter thought he'd have a big story if he could get Mr. Ford's reactions now that the bill had become law.

I took this reporter to see Mr. Ford. The reporter told him what he wanted, and Mr. Ford nodded. Then he went on to talk for half an hour about a new million-dollar generator he was buying for the Rouge. When he finished with that discourse, the interview was over.

"Hell," the reporter said to me later, "that's no story."

But the reporter was stupid. Talking about something else was Mr. Ford's way of showing how seriously he took the Wagner Act. It didn't exist, for him.

As the CIO's efforts to unionize the Ford Motor Company became more intense, there were accusations that I was spying on the union. The accusations were unfounded. Actually, it was a waste of time to send my men into the union to get information. We didn't have to.

There were always a certain number of men who thought the union wouldn't win out and wanted to stay in good with us. After every meeting, these men would come to us and tell us what had transpired. We were good listeners. When it seemed advisable, I would take one of these informants up to lunch with me in the executives' dining room and let him talk all he wanted. With so many coming to see me that way, I could always cross-check their stories, and would know when they were telling the truth.

At least one bit of comedy relief came into the whole union situation.

The union was by now broadcasting over a little radio station in Detroit in the Maccabee Building, a station so small that few people had ever heard of it, and fewer still could get it on their radios.

One day Benson Ford, Edsel's second son, came to see me and asked if I ever listened to these broadcasts. "I get a kick out of listening to those fellows," he said. "You should listen to them sometime."

I would have liked to, I told him, but I never could get the station. "How the devil do you get it?" I asked.

"Oh, that's easy," Benson said. "You just push one of the selector buttons on a Ford car radio, and you get the union station."

Evidently we had some union man in the radio department who was setting one of the selector buttons to the Maccabee station on every radio that went out of there. Since all of our employees drove Ford cars, when the union broadcasted it spoke to almost every man in the plant.

I went over to see Ray Rausch, a production executive working under Sorensen. He didn't know a thing about it. We changed that situation, though, in a hurry.

Incidentally, none of this stuff ever bothered Sorensen. If it had been up to him, he'd never have bothered to change the buttons.

After a while, of course, we got pretty well involved with the NLRB. In time, they had cases against us both in Detroit and in our eight assembly plants. None of these, as I recall, went in our favor. By then our reputation with labor was so bad that Mr. Ford's curt remarks on the subject preceded our attorneys into the courtroom.

After a number of hearings Lou and Edsel decided that we needed a good constitutional attorney if we were going to appeal any of this. Mr. Ford agreed, and Edsel, Lou, and Longely decided to hire the firm of Frederick H. Wood of New York.

So Wood was engaged—for an extremely high fee.

I don't know why this firm was hired, because they didn't improve on anything Colombo had done. They didn't get any farther in the courts than Colombo got—which was nowhere.

As the conflict with the UAW increased, we'd get calls from other plants—mostly our suppliers—that were already unionized. They'd say, "You're not going to give in to those birds, are you?"

After a while this encouragement from people who had themselves already "given in" both amused and annoyed me. I pointed out to Mr. Ford that what they were saying was: "Let's you and him have a fight."

One day Clarence Avery of Murray Body called me up at a moment when Mr. Ford was in my office and asked if we knew of a good man to handle their union troubles. I relayed this request to Mr. Ford, and he said, "Tell 'em, if they can't lick 'em, join 'em. That's what we're going to do."

What Mr. Ford had in mind is best shown by an incident that happened some years earlier.

At one time the Ku Klux Klan was organized within the Ford Motor Company, as it was in other plants. It began getting pretty ugly. Mr. Ford and I got worried, and decided it ought to be broken up.

I went to one of my clerks and said, "How would you like to join the Klan?"

"Why," he said, "I'm a Catholic."

"That's fine. All the better," I said. "I think it would be a good idea if you and some of the other boys joined up."

He got the point, then, and he and quite a few other men close to me joined.

Well, they just disrupted the devil out of the thing. They had more fun—after a while they got the Klansmen so mixed up they didn't know which way to turn. The men I'd sent in got to accusing each other of being spies, and this only added to the uproar. Finally, one of the men who had worked his way up to be treasurer ran off with most of the dough.

That ended the Klan in the Ford plant. After that, if you went up to someone and said "Klan," you'd either get no attention at all or a sock in the nose. The whole thing was broken up by ridicule.

It was Mr. Ford's idea that the same sort of thing might be pulled with the union. But it didn't work out that way. They were too well supported from outside.

Chapter 16

MEANWHILE, other important events were occurring within the Ford Motor Company. For in the thirties Mr. Ford's name became linked with Nazi activities, both at home and abroad. While I cannot claim to know everything that went on in this area, I was involved in some of it.

It was my role to try to protect Mr. Ford against himself. I knew that all this stuff was hurting Mr. Ford personally, and was hurting the company. When I could prevent some rash act on his part, I did so. When I could not prevent it, I tried to make Mr. Ford's plunge off the deep end as restrained as possible. And when I could not do even that much, then I tried to patch up the pieces.

This task was not made any easier by the fact that, in his later years, Mr. Ford believed he had divine guidance. I don't know how many times I reproved him for some rash act, and he pointed to his head and said, "I'm guided, Harry, I'm guided."

In 1938 two German engineers brought the Grand Cross of the German Eagle to Liebold for Mr. Ford. Mr. Ford accepted the decoration. By this time Hitler's name was odious to a great section of the American people.

I felt at the time that Liebold had roped Mr. Ford into this thing, and, knowing the probable results, I jumped on Liebold for his part in it. However, he would only give me an abrupt defense of his action, pointing to our plant in Cologne, Germany, and claiming it was necessary to accept the medal for business reasons.

I never felt the medal meant much to Mr. Ford. He ridiculed it to me and to other people. I believe his acceptance was just one of those many things he did out of mulishness and ignorance, and a failure to realize the consequences of his actions. I think, too, that one of his motives in accepting the medal was that he thought it would get F.D.R. mad. Mr. Ford hated Roosevelt.

The newspapers published a picture of Mr. Ford, standing with the two Germans, and with the medal hanging around his neck. When I reproved Mr. Ford for having this picture taken, he said, "Why, I never had my picture taken with that medal. That was faked."

I told this to Liebold, and said I was going to raise cain with the papers. He said dryly, "I wouldn't do that."

I didn't quite know what to believe then. I called one of our company photographers and said that I was about to bet someone that the photograph of Mr. Ford wearing the medal was faked.

"Don't bet any money on it," he advised. "I took that picture, and I didn't fake it."

After the publication of this picture and the accompanying news story, there was a marked decline in our sales. Officials of the company soon became convinced that there was an active and effective boycott against us. We tried to convince Mr. Ford of this, but he stubbornly refused to believe us, however many sales charts and figures we showed him.

Exactly what Mr. Ford's attitudes were toward Hitler has long been a matter of controversy. Perhaps, having been as close to him as I was, I can help clarify that question.

A number of the ingredients of Mr. Ford's personality went into making up his total attitude. When I told the story of how Mr. Ford helped the ousted mayor of Detroit, Charles Bowles, I mentioned how, when everyone jumped on some fellow, Mr. Ford was all for that man. Some of that went into Mr. Ford's attitude toward Hitler.

I also mentioned Mr. Ford's general mulishness—and some of that went into it, too.

Also, Mr. Ford was strongly anti-British. I believe this attitude dated from a visit he had from Winston Churchill. During their talk together, Mr. Ford told Churchill he thought most of England's problems could be solved if farming were encouraged in England, so she could more nearly produce her own food. Churchill ridiculed this suggestion, and I think Mr. Ford never got over that.

Finally, Mr. Ford was definitely pro-German. He considered that the German people were clean, thrifty, hard-working, and technologically advanced, and he admired them for that. During the early years of Hitler's rise to power, Mr. Ford just stubbornly refused to believe the stories of Nazi violence and brutality, and put it all down to propaganda.

Now, if you add all these ingredients together and stir briskly, you get Mr. Ford's attitude toward Hitler—up to a certain point.

Mr. Ford was convinced that once Hitler

got the Polish Corridor, he would seek no further territorial gain, but would turn to building up Germany. When, however, Hitler went right ahead with his conquests, Mr. Ford changed.

He said to me then, "Well, by God, we're through with him. He's just power-drunk, like all the rest of them."

A good way to show Mr. Ford's position on the international scene in the late thirties would be to tell three anecdotes that are somewhat related.

Shortly after the war had begun in Europe, an Englishman came to see Mr. Ford in the hope of persuading him to help bring a lot of English children over here to protect them from bombings. There was quite a movement on then to do that sort of thing.

Mr. Ford refused to help. He told the Englishman that if the British were really concerned about their children, all they had to do was to put all their children in one part of England and notify the world; the Germans would never dare bomb that part of the country.

Mr. Ford jibed at the man, saying, "Do you really want to get rid of Hitler? I'll send Harry over there with six of his men. They'll get rid of Hitler for you in no time."

Then he went on to inform his visitor that in his opinion England had started the war at Munich. Referring to Chamberlain and the Munich pact, he said, "Why did you send that fool with the umbrella over there to talk to Hitler? Why didn't you send a man?"

The second anecdote involves one of the German engineers who brought the medal to Mr. Ford.

This man came to see Mr. Ford a second time, and began orating at him in his rich Germanic accent.

"Mr. Ford," the man said, "you can do a great deal for Germany, and the Fuehrer . . . the Fuehrer . . ."

"Aw," Mr. Ford broke in, "the fewer the better," and he got up and walked out of my office.

The German was stunned. Incapable of understanding Mr. Ford's pun, he asked me what Mr. Ford had meant, but I knew it was hopeless to try to explain, and I just stood there laughing.

The third story I have in mind occurred when Eric Johnston came back from his trip to Russia.

Johnston came back from Russia with a message from Stalin to Mr. Ford, and he came out to the Rouge to deliver it in person. Johnston, young Henry, and Mr. Ford met in my office.

Johnston, who is quite a talker, gave Mr. Ford a lengthy discourse on his trip, the gist of which was a plea for understanding of the Soviet Union's war role.

Mr. Ford listened in silence for ten or fifteen minutes. Then suddenly he got up from the chair.

"Aw, you're nothing but a preacher," Mr. Ford said, and again walked out of my office.

As the reader will see, and as I have said before, Mr. Ford was scarcely a predictable or consistent man.

Those qualities didn't help much, either, when it came to protecting him against his own follies.

At one time John Bugas, then head of the Detroit FBI office, informed me that the manager of one of our foreign plants was bringing information to subversive groups in this country on his frequent trips. Bugas claimed that the FBI had pictures of him talking with known spies in this country.

I told some of this to Mr. Ford.

Shortly after that I got a call from Mr. Ford. He said, "——— is in Liebold's office. Do you want to come over?"

By then I was afraid the man probably had a tail on him. I checked, but couldn't find anything suspicious, so I went to Liebold's office.

Mr. Ford talked to the fellow a short while. Then, as he got up to leave, he said,

"Listen, you're nothing but a goddamn spy."

And that's as much as the man ever got out of Mr. Ford—at least while I was with him.

The FBI came into the picture on still another occasion—but in a very disappointing way.

At the start of the war I was informed by one of our Mexican distributors that a certain branch manager was so pro-Nazi that he was causing a lot of comment in and around Mexico City. The papers were already printing so much stuff that was harmful to us that I knew they'd lay this to Mr. Ford, although he didn't even know the man. So I wanted to get that fellow out—fast.

I checked on my information, which was easy to do, and found that it was correct. I informed Edsel of the matter. He said, "If that's so, we ought to throw him out. But first we should get the FBI to countercheck."

While having lunch with John Bugas, I gave him my information about the branch manager. I said we'd fire the fellow, if the FBI would give us a letter of their findings in the matter.

Apparently Bugas, who didn't know I'd spoken to Edsel, arrived at the conclusion that I was trying to get rid of a man behind Edsel's back. He conveyed my request to J. Edgar Hoover, and in all probability added his conjecture as to my motives.

I conclude all this, because Hoover at once wrote Edsel. He informed Edsel of my request to check on the branch manager (which of course was Edsel's idea in the first place) and asked Edsel if he was aware of what was going on in the plant.

Edsel brought me Hoover's letter, and I read it. I was astounded. We had assumed that the FBI would simply investigate the man and send us a report that he was subversive, and then we could just put the fellow out on the basis of this report and have no trouble over it.

Edsel said, sheepishly, "It looks like it backfired."

"Yes, it does," I said.

"Well," Edsel said, "Hoover is a friend of mine, so just get rid of the branch manager anyway—but keep me out of it."

So I got rid of the fellow—without the help of the FBI.

In 1939, about a month after leaving Yale, young Henry, Edsel's eldest son and now president of the Ford Motor Company, married Anne McDonnell, daughter of an Eastern broker. Before their marriage, Henry became a Catholic convert under the tutelage of Monsignor Fulton J. Sheen.

If the reader bears in mind the senior Ford's prejudices, it can be understood that this was no small event in the Ford family.

After that, even when Henry came into the company a few years later, Mr. Ford always considered that Henry was under the "influence" of Sheen. When Henry did something Mr. Ford didn't like, Mr. Ford would say to me, "Don't pay any attention to what he says any more. Someone's talking to him."

But the big blow-up came when it became public knowledge that Henry was changing his religion. Walter Winchell intimated on his radio program that Mr. Ford himself might be the next convert. When he heard that, Mr. Ford, who disliked Winchell anyway, just about went crazy.

He called me up at home at ten o'clock at night and wanted me to bring Dave Willkie of the Associated Press down to the office right away to meet him, so that he could make a statement to the press. I managed to talk Mr. Ford out of this, convincing him that such a move would only please Winchell.

Mr. Ford didn't get over it that easily, though. He wanted to sue Winchell, and went around for days raging against Winchell to everyone he met.

Finally Mr. Ford worked out his own answer to Winchell. One day when he was in my office he suddenly asked me, "How far can I go in Masonry?"

He had been a Mason for years, but had never paid much attention to it. If his dues hadn't been paid out of the office, he might have not even have kept up his membership.

I said, "I don't know. How far are you?" And then it occurred to me that Harry S. Toy was in my outer office waiting to see me, and I added, "There's a thirty-third-degree Mason right outside. Do you want to talk to him?"

Mr. Ford said he certainly did, and Toy came in. Mr. Ford asked him how far he could go, and Toy said, "Why, you can go all the way."

Mr. Ford didn't want to travel to Boston, which is where the thirty-third degree is given, so Boston came to him. All the people who were necessary to do the job just came to Detroit.

Of course, it took some days to run Mr. Ford through all the degrees, but he went through it cheerfully. Every time he was about to go downtown to take some more degrees, Mr. Ford would say to me, "I'm going down to take care of Winchell."

Actually, Winchell had a reason for saying what he did. Mr. Ford had a regular "line" he used with every clergyman he met. Whatever the man's denomination, Mr. Ford would always say to him at some point, "Well, now, you've got the best religion in the world."

I know that Mr. Ford said that to Sheen when he met him, and I suppose it got to Winchell somehow.

In any event, once Mr. Ford was a thirty-third-degree Mason, he felt better about the whole thing.

Chapter 17

THOSE were hectic times in the Ford Motor Company, those prewar years. Not only were we getting a bad press on the Nazi stuff, but we became involved as well with some people at home who did us no good. Perhaps this is an appropriate place to tell that part of the story.

Sometime in 1939, Elizabeth Dilling, author of *Red Network,* went several times to Mr. Ford's office to try to talk with him. Finally, Frank Campsall asked me to see her.

I had a talk with her, and she told me she needed $5,000. I reported this to Mr. Ford, and he said, "Well, give it to her."

I was afraid, though, that if I gave her this money in a lump sum she would go out and tell the world that Mr. Ford was in back of her. So instead, I put her on the payroll for six months.

This was just about the time when Kipke was running for a place on the University of Michigan's Board of Regents. To give Liz something to do, I sent her to Ann Arbor to get some dope on Communists there, thinking she might provide some ammunition for Kipke to use against the faculty. Actually, she did a pretty good job, although probably not much better than any competent newspaperman might have done. She spent about a month at this.

I was far from comfortable with Liz around. When she walked into a room her first act was to go looking behind all the pictures on the wall, searching for dictaphone mikes. Sometimes, when she got into my office, she made terrible scenes, and I could hardly get her out.

When she was finished with the job for Kipke, Liz confided to me that she needed that $5,000 in order to go to Europe, where she wanted to gather material to write a book about the Communists there. I saw this was just the "out" I wanted, so I told her she wouldn't have to wait, I'd give her the balance of the money Mr. Ford had agreed to pay her right then.

I did, and off to Europe she went.

I didn't tell Mr. Ford about it until she was safely on the boat.

In that same year—1939—we became involved with Gerald L. K. Smith.

I had never heard of Smith when some big businessmen in Detroit came to see me about him. I was told that Smith was just up from Kansas and that he was an eloquent spokesman against the CIO. At this time, I wish to make it clear, Smith was not yet peddling his "nationalism" and his bigotry. He only turned to that later—to get other suckers.

These businessmen suggested that we help Smith combat the union movement. At first I felt disinclined to do this, for I was afraid it would only give the union more publicity. It was an established policy with Mr. Ford and me never to refer to fellows like Reuther and Frankensteen by name, but only as "union bosses." But finally I agreed to join with them in helping Smith.

However, I quickly became disillusioned with Smith, and resolved that I would have nothing more to do with him.

The way Smith got into people's confidence was by using his wife. Smith had a beautiful and gracious wife, and he always used her to run interference for him. When he came to see you for the first time he would bring his wife along, and she made such a good impression you thought there must be something to the fellow.

If he came to see you again—perhaps

under more difficult circumstances—he brought her along too. He knew you wouldn't want to throw him out on his ear while his wife was present.

Soon Smith came to ask for more money. I refused.

However, like all fellows of his sort, once he got his hooks into you, it wasn't so easy to shake him off. I kept giving him the run-around, but he kept right on trying.

He tried to get money from Mr. Ford, but I prevented this, though Mr. Ford wanted to give him some.

One day Smith called me up in a panic. He said that Giles Cavanaugh, Collector of Internal Revenue for Detroit, was checking on his income tax. Smith asked me to try to persuade Giles to call off his dogs.

I bluntly refused to help Smith, but that wasn't enough to get him out of my hair.

Smith decided he wanted to run for governor of Michigan. Of course, he couldn't have been elected dog catcher, but he took his own political ambitions very seriously.

By this time Slim Lindbergh was a consultant for the Ford Motor Company. Smith wanted to get Lindbergh to help him run for governor. I advised Slim against this, so Slim just avoided seeing him. Smith, however, was not inclined to give up.

One day Slim came out to my ranch house on the Huron River to go riding with me. Smith found out about this by calling the Dearborn Inn and pretending that his call was expected.

Next Smith called up Service at Highland Park, knowing they wouldn't check with me, and told them I wanted to see him. He was a great bluffer, and this time it worked. They sent him a car and driver.

Smith then had himself driven out to my ranch house—his wife beside him for protection, of course. Slim and I had just got our horses saddled and were down by the stables. I was astounded to see Smith. I couldn't understand how he'd found me.

I told Slim, "Now, don't you say a word, I'll handle this," and he just sat there on his horse without speaking. I took the driver aside, and he told me how Smith had got out.

I then went into the stable and called up I. A. Capizzi, by then Ford general counsel, and asked him to come over. When he showed up, I turned the whole thing over to him. Then Slim and I rode away.

Incredibly, Smith still didn't give up.

He began calling my office, trying to see me. In desperation he began annoying my clerk.

I called Smith on the telephone. I said, "If you ever bother me again or any of my clerks, I'll run you right out of Detroit, and I won't need any help to do it, either."

And that, happily, was the last I heard from him.

However, it was not the last of my troubles with that kind of people. Next to come into my life was William Dudley Pelley, leader of the Silver Shirts.

Bill Cameron brought to Mr. Ford a pamphlet written by Pelley in which Mr. Ford was named and some of his views quoted. Cameron said he thought Pelley was a pretty good man, and perhaps we ought to get in touch with him. Mr. Ford was favorably impressed by the way Pelley had quoted him, and asked me to find out what I could about the man.

I sent Frank Holland and Jim Brady down to South Carolina to find out what kind of man Pelley was.

To get Pelley's confidence, Holland and Brady bought a batch of Pelley's newspapers, assuring him they wanted them for distribution at the Rouge. Delighted, Pelley gave them any information they wanted. Later they secretly burned the newspapers.

When they got back, Holland wrote out a report for me.

"The first afternoon," Holland began, "we couldn't get to see Pelley, because he was up walking with God."

Holland then went on to report that he understood Pelley to say that every afternoon he went up into the sky to see God, got his instructions, and then came back to earth and did whatever God wanted him to.

When Cameron read this report, he tried to put another face on the matter. By this time he hated me so much that he wouldn't even speak to me on the phone, but he said to Mr. Ford, "Why, all the man meant was that he was deeply religious, and he communed with God on occasion."

Probably encouraged by Cameron, Pelley came up North to try to see Mr. Ford. But I wouldn't let Pelley in to see him, and finally he just gave up and went away.

This chapter wouldn't be complete without mention of Fritz Kuhn's association with the Ford Motor Company.

Kuhn was hired and put on the payroll—but not by me. Kuhn, you may recall, was known as the American Fuehrer.

Of course, the papers all knew of Kuhn's employment by the company, and as his reputation became more and more odious to the American people, just so much more harm did he do the company and Mr. Ford's reputation.

I clearly understood all this, and for quite some time watched for a chance to fire Kuhn in a manner that would have no kickbacks. I wanted to be able to get rid of him under circumstances where he couldn't cry "persecution," and also where he would be unable to bring us any further bad publicity.

My chance came in 1937. Kuhn stopped an elevator between floors in the Henry Ford Hospital and made unwelcome advances to a nurse. The moment I learned of this, I bounced him.

It came off as quietly as I could have hoped. Kuhn had no desire to talk about it at all. We notified the FBI that we had fired him, but said nothing to the public.

Liebold was furious. He said to me, "Why don't you fire him for the *real* reason you're firing him?"

But Liebold knew perfectly well I didn't want to fire Kuhn for political reasons. I liked it fine just the way it was.

Two years later, in 1939, Fritz Kuhn was convicted on a charge of having embezzled Bund funds.

Chapter 18

LOU COLOMBO left us in 1940, and he did so in anger because of a political fracas we became involved in. Under the circumstances we were afraid he would raise the race issue against us. So Mr. Ford said, "Well, we'll just hire another Italian."

A lot of people, including a Detroit Free Press representative, urged us to hire I. A. Capizzi. I had never known him socially before, nor had Mr. Ford. But on the strength of all the recommendations, Capizzi was hired through Longely, Edsel's attorney.

The original idea was that Capizzi was to work under Longely. However, Mr. Ford liked Capizzi immediately, and he shortly told Capizzi that he couldn't work for

Longely and him too. So Capizzi was encouraged to set up his own office.

The union situation was getting more and more critical, and Cappy stepped into a job that was destined to keep him pretty busy. For one thing, we were carrying an NLRB case brought against us in Detroit up to the Court of Appeals, and when we lost it there we carried it to the Supreme Court on the grounds that the NLRB was unconstitutional.

As the case approached a hearing in the Supreme Court, Mr. Ford asked me to find out what Justice Felix Frankfurter thought of us. Mr. Ford was worried.

I sent Frank Nolan, whom I have already mentioned as a company attorney, to Washington. Nolan was a good friend of the late Justice Murphy, and could get into those circles.

After a few days Nolan called me up. "How do you want it, boss? Do you want it straight, word for word, just like I got it?"

"I want it straight," I said, "just like you got it."

"Well," Nolan reported, "Frankfurter said, 'It's about time Mr. Ford changed his attitude toward labor.' "

I gave this information to Mr. Ford, and asked him if we shouldn't stop the appeal.

"Why, no," he said. "It's worth going ahead just to find out." And then he repeated what he had told Governor Fred Green at the time of the incident over Butzel. "However a man gets on the bench, he changes. It does something to him. Maybe we'll be all right."

As it turned out, we lost. The Supreme Court refused to review the case.

However, the only thing we got out of it came from Frankfurter. The NLRB had directed us to stop publishing the Ford Almanac, which dealt with the union and other things. Frankfurter ruled that our rights of free speech were involved and that we could continue to publish the magazine.

And thereby hangs a long story.

It has been said that "nobody loves a fat man," but Mr. Ford went further than that; he seemed to have a deep personal aversion to corpulent people.

I decided that we should publish the Ford Almanac again, and that the best editor would be Jimmy Sweinhart, who worked for Cameron, and who had edited it previously. However, Mr. Ford disliked Sweinhart. It was frequently Jimmy's job to look after Cameron, and Mr. Ford said of him, "Aw, he's just Cameron's nursemaid." Also, Jimmy was a big, heavy man. I thought, though, that I might persuade Mr. Ford to see it my way, and with that in mind I got both men in my office together.

But before I can tell you what happened, I must tell another story.

Fred Allison, a big three-hundred-pound man, was one of the real pioneers of the Ford Motor Company. I believe he was the developer of the AC spark plug and many other important things. I believe, too, that he may have developed the Allison engine.

Allison was an independent man who did pretty much as he pleased, however, and didn't show too much respect for Mr. Ford's wishes. Although he was well aware of Mr. Ford's attitude toward smoking and drinking on the job, Allison did both openly. When Mr. Ford tried to curb him, Allison resigned from the company.

Many years passed, and the time came when Allison was out of a job. He went to see Mr. Ford, and Frank Campsall sent him to me.

Allison eased his bulk into a chair in my office and said, "I've got to have a job."

"Mr. Allison," I told him, "I haven't got anything in your bracket."

Allison was about to answer when Mr. Ford came into my office. He said hello to Mr. Allison, and then I told Mr. Ford why Allison had come to see us.

"Well," Mr. Ford said, "he's worth about six dollars a day." That was his favorite insult.

Allison said, "That's all right. I need a job, period."

Mr. Ford looked up and down Allison's vast bulk, and after a moment said, "You go away and take off fifty pounds, and then come back and see us."

"All right," Allison said.

Allison went away, and sure enough, some time later he came back again, looking thinner. Mr. Ford came over to see him, chatted a while, looked him up and down again, and said, "By God, you're still too fat. You go away and lose another fifty pounds. Then come back again."

Allison went away. Still later, he came to see me once more. He was now much thinner. I decided not to let Mr. Ford see him again. I sent Allison over to a substation to work as an electrician.

However, as I have already explained, it was just about impossible to keep from Mr. Ford anything that was going on around there. One day he came into my office and said, "I hear Fred Allison is working here."

"Yes, he is," I said.

"Well, let's go over and see him."

We went over to the substation, where we found Allison at work. Nature had given him a big frame, and had never intended him to be so thin; he looked like a dead man.

Mr. Ford and Allison chatted a while, and then Mr. Ford said, "Looks like you're a little thinner, Fred."

Allison took hold of his clothes and held

them out at the waist, to show how much weight he had lost.

"By God, you're still too fat," Mr. Ford said. "You just go out and buy a new suit of clothes that fits you. You're leaving those clothes on because you want to eat your way back into them."

We left, then, and when we got outside Mr. Ford said proudly, "He can't fool me."

I said, "I think this is the most hideous thing I ever heard of."

"You let fellows pull the wool over your eyes," Mr. Ford said, "but they can't pull it over mine. That's one trouble with you, Harry—you're too soft."

Allison worked in the substation seven years, and then died.

Well, as I said, Jimmy Sweinhart was a huge man. Mr. Ford looked him up and down and said, "You go away and take off fifty pounds and then come back and see us."

I wasn't going to let that whole thing happen all over again, and I said, "Now, Mr. Ford, he's a big-boned man. He can't lose that much weight. Don't you remember how Fred Allison died?"

Mr. Ford looked at me and said, "Oh, did he die?"

By the spring of 1941 our relations with the CIO had reached a critical point, and in early April the plant was closed down.

I put it that way because I will never agree that the Ford plant was struck. Only our steel mill really went on strike. The Rouge was closed down because the plant wasn't given the protection it should have had. The union simply blocked all the roads and wouldn't let men into the plant.

As an example of how far the governor went to help the thing along, Don Leonard, now state commissioner of police, who had been so helpful in disturbances before, was sent on a Southern trip to get him away. He called me from New Orleans to tell me he was there.

With the plant closed down, an astounding thing happened. White-collar executives disappeared. They just seemed to evaporate into thin air. Ray Rausch took over the plant and stayed there night and day. Besides him, Mr. Ford and I were the only two executives who went into the plant during the whole strike.

While he never showed up at the plant, Edsel did keep in touch with us by telephone. He was in Florida when the plant was closed down, and when he learned what had happened he went up to his country estate north of Ann Arbor. From there he kept in touch with me by phone.

Mr. Ford wanted to fight the thing out. He told me to arm everyone we had in the plant, and use tear gas if necessary. I felt the same way Mr. Ford did.

Edsel, however, became extremely alarmed when he learned of our plans, and insisted that we give up any such ideas. Though Mr. Ford and Edsel were far apart on this, Mr. Ford gave in to Edsel's wishes. I don't think the CIO would have won out if it hadn't been for Edsel's attitude.

Since we couldn't fight it out, there was nothing to do but try to settle it. At first, though, Mr. Ford didn't want to settle. He wanted to sell the plant.

Mr. Ford became convinced by certain people close to him that "the Jews" were behind the union, and that the whole thing was a "Jewish plot" against him. So thoroughly was Mr. Ford convinced of this that he preferred selling the plant to "giving in."

I—and I alone—convinced Mr. Ford of his mistake.

I said to him, "Will you go with me and meet the biggest Jew in the union?"

"Sure I will," he said.

"Well, I want you to meet Phil Murray."

"Why, he's no Jew!" Mr. Ford exclaimed.

"No," I said, "and neither is Homer Martin or John L. Lewis or Bill Green or any of the rest of them."

In time, Mr. Ford saw how ridiculous his belief had been, and then—and only then—was he ready to enter into discussions with the union.

I believe this story shows that you could

always take issue with Mr. Ford. And I believe if others had had the courage to take issue with him, he might have been a different man.

We began discussions with the union at the Dearborn Inn. When we didn't seem to make much progress, both Madame Perkins, then Secretary of Labor, and the President, who were in touch with the situation, insisted that Governor Van Waggoner see Mr. Ford.

I took Van Waggoner out to the Residence. We walked into the house and out on the sun porch, where we found Mr. Ford reading a paper.

Mr. Ford took his reading glasses off, looked at Van Waggoner, and said, "Well, you've got a plant—what are you going to do with it?"

Van Waggoner became very agitated and seemed incapable of coherent speech. He repeated a meaningless phrase over and over.

When it became evident that the governor could say nothing more, I led him out.

Later Mr. Ford told me, "I'd give him six dollars a day—but not any more."

In the end, we agreed to a government-supervised election. The election was held on May 21, 1941, and the CIO won it.

A month later we signed a contract with the UAW that included everything they had asked for and a lot they hadn't asked for, and was far in advance of anything then in force in the industry. Countless biographers have tried to explain how this came about, and all went far wide of the mark. I want to tell, now, how that really happened.

Chapter 19

AT FIRST, negotiations for a contract didn't go too well. In fact, I recall one meeting in which Phil Murray threatened to walk out on the whole thing, and negotiations were kept going only by an amusing move on the part of Frank Nolan.

Nolan had studied law while working for me. Because of his wit and his talent as an actor, it took Frank eight years to get his degree; he was invited to every party in town, and couldn't say no. When he did finally get his degree, he sent out announcement cards, as every young lawyer does, but Nolan's read: "Francis Aloysius Nolan, attorney at last." He became associate general counsel under Capizzi.

At the negotiation meeting of which I

spoke, there were present Mr. Ford, Capizzi, Harry Mack, and Nolan for the company, and Phil Murray, T. J. Thomas, and Allen Haywood for the union. As I said, matters reached a point where Murray threatened to walk out.

There was a moment's silence, and then Nolan said, "Phil, let's go into the next room, get down on our knees, and say the Litany of the Blessed Virgin, and let the rest of these Protestants go to hell."

They did, too. Nolan and Murray went into the next room and prayed together. When they came out everyone was in a better frame of mind, and the meeting went on.

The first and biggest stumbling block to get over was the question of wages. Mr. Ford insisted to me that we were paying higher wages than any of the other automobile manufacturers.

"No, we're not," I'd say, but he'd go right on insisting.

Finally I sent to the accounting department for the figures, and showed him conclusively that our scales were below the others.

Mr. Ford then said to me, "By God, Harry, we ought to be able to pay more than General Motors or Chrysler. They've got stockholders to settle with, and we've only got the family."

With that, I went to CIO National Headquarters in Washington and made an offer to raise Ford wages above those paid by Chrysler and General Motors. When I made this announcement, all the union men looked thunderstruck and didn't say a word, so I said, "I'll go back to Detroit and let you think it over."

At that, I heard one of the lawyers present whisper to a union official, "Grab him. Don't let him out of the room."

In addition to this, we offered the union the "closed shop"—that is, we agreed that every employee would have to join the union in order to keep his job. And we offered the "check-off," which meant that we would deduct union dues from the wages of each Ford employee, and remit these to the union treasurer.

The union had not asked for either provision. They had won them nowhere else in the industry. Why did Mr. Ford offer them?

Mr. Ford told me, "Give 'em everything—it won't work." He then explained that he felt if we gave the union just a little, then they'd be right back at us for more. But if we gave them "everything," he thought, then they would fall to fighting and bickering among themselves. The way he saw it, it was a case of "enough rope."

In addition, Mr. Ford rather liked the idea of the check-off. He said to me, "That will make us their bankers, won't it? Then they can't get along without us. They'll need us just as bad as we need them."

And that's the story of Mr. Ford's motives in making the offer to the union that he did, here told for the first time. When we reached full agreement, I went to Washington and signed my name to the first contract between the UAW and the Ford Motor Company late in June.

Actually, the negotiations with the union went on for some time after the signing of the contract, since a number of questions were left open. Among other things, these post-contract negotiations led to the departure from the company of Crawford, Edsel's assistant. Here is how it happened:

The union negotiators began demanding that they talk with some of our officials; so little did they understand the administrative setup in the company that they felt they were getting the run-around because they weren't meeting with the fellows who had the titles. Consequently, I told both Craig, who was secretary and treasurer of the company, and Crawford, who was Edsel's assistant, that they would have to meet with the union men.

"What?" both men said. "Do you mean we have to talk with those ———?"

This expression of contempt made me angry. My own contempt was for the white-collar officials up in mahogany row. I had respect for many of the workmen in the plant; there were plenty of intelligent men among them who could have taken over these officials' jobs without the least bit of trouble. But there were no officials—with the exception of Sorensen, who shunned mahogany row and could build a car from the wheels up—who, in turn, were capable of doing the jobs these workmen were doing.

I arranged a meeting. The union negotiators got to my office first. I was still angry about Craig's and Crawford's attitude, and so, while we waited for these men and other officials to show up, I encouraged the union men to talk up to them. "Now," I said, "you've been demanding to see these fellows; be sure you give them hell."

When Craig and Crawford and the other officials showed up, the union men lit right into them. The officials looked among themselves for a leader, and found none. I just let them flounder along.

After this fiasco, Crawford told Edsel he didn't like the way I was handling the union. He told Edsel that I was "giving away the plant."

For all their antagonism, Edsel never kept anything from his father. He told Mr. Ford what he had heard from Crawford.

Mr. Ford came to me and told me about the whole thing. He said, "Harry, you get rid of Crawford."

Mr. Ford was worried about Edsel's

health. Edsel was already ill with the cancer that was to kill him within two years, and Mr. Ford wanted Crawford out because he felt Crawford was upsetting Edsel. He blamed Edsel's ill-health on the "ear-piddlers" around him.

I told Mr. Ford I didn't want to do that. After all, relations between Edsel and myself were never of the best, and if I fired Edsel's assistant, matters between us would be irreparable.

The upshot of it was that Sorensen took Crawford over to the plant with him, and later reported to me that he was getting on with Crawford very well. "I'm not having any trouble with this fellow," Sorensen said. "I think he can work with me all right."

I told Mr. Ford what Sorensen had said, but he wasn't satisfied. He said, "Aw, I don't want him around here. I'll have a talk with Charlie myself."

Which meant, of course, that Crawford left.

I have been speaking about events that occurred in the early part of 1941. By that time, as we now know, World War II was almost at our doorstep, but few Americans realized then how close it really was.

Mr. and Mrs. Ford were both members of the America First Committee. I was worried about that; I didn't like seeing their names there. But there wasn't much I could do about it.

In the summer of 1941 Slim Lindbergh came to see Mr. Ford. The visit was noted in the papers, but the reason for the visit was not known. Lindbergh asked Mr. Ford to make a contribution to the America First Committee, to be sent to General Wood in Chicago.

Mr. Ford told me to take $10,000 in currency out of the kitty and send it to Chicago.

I didn't think this was a wise thing to do, and decided to get the money from the company paymaster.

As it happens, Arthur Hatch, then our Chicago zone manager, was traveling from New York to Chicago with some Navy people in a private car lent for the trip by the Michigan Central. At the time, his car was in Detroit. So I had the paymaster take the money to Hatch to take to Chicago for us.

My uneasiness about Mr. Ford's connection with America First was increased by this whole deal. So as soon as the money was delivered to Hatch, I began hammering at Mr. Ford, telling him we really needed that money in Chicago for entertainment, and that it should be used for that purpose. Finally I convinced him.

The money was never delivered to Mr. Wood.

Shortly after this, Mr. Ford, just to please Edsel, came pretty close to making a gesture in support of our foreign policy—but didn't, quite.

Secretary of the Navy Frank Knox called me on the phone from Washington and suggested that Mr. Ford make a radio speech, urging the American people to get behind the President's foreign policy.

I told Mr. Ford about Knox's suggestion and Mr. Ford promptly said he wouldn't do it.

I said, "We're going into war pretty soon, and I think you ought to do it."

Mr. Ford thought this over, and then said, "I wonder what Edsel would think about it."

I called up Edsel at Colorado Springs, where he was vacationing. Edsel was very friendly with F.D.R., and approved of most of what he did. Edsel said, "I think it would be wonderful—but I'll believe it when I hear it."

I reported to Mr. Ford simply that Edsel thought it would be wonderful.

"All right," Mr. Ford said. "I'll do it."

So plans for the broadcast were made. They wired up Mr. Ford's office for the purpose, and then he decided he'd rather talk from home, and they wired up the Residence.

But two days before the broadcast was to have taken place, Mr. Ford backed out.

Some people close to him called the project "a lot ot tomfoolery," and talked him out of it.

But all these events notwithstanding, I suppose the single most important occurrence at the Ford Motor Company at this time was the beginning of Willow Run.

Willow Run was Sorensen's idea, and he sold Mr. Ford on it sometime in 1940. Cameron's publicity department immediately went overboard on the thing, and promised the public that we would soon be turning out a thousand fighter planes a day—a wild promise that was later to be recalled to our discomfort.

Sorensen drew up the rough production layout for the building, and it was then designed by the late Albert Kahn, who was the architect for every big building Mr. Ford ever put up. Construction began in January 1941.

Willow Run was to manufacture the Liberator B-24, a four-engined bomber designed and originally produced by Consolidated Vultee Aircraft. The largest and the most elaborately tooled aircraft plant in the world, it was a mile long and a quarter-mile wide, and was built at government expense for about $65,000,000. To accomplish the manufacture of such an airplane under one roof, seventy subassembly lines were set up, and earliest estimates for manpower needs were for 100,000 aircraft workers.

Willow Run was built in a little over a year. It began production in May 1942.

Meanwhile, we had already erected at the Rouge a plant for manufacturing airplane engines. We called it the Pratt & Whitney Building, since we made Pratt & Whitney's engines. About this time something happened there that will perhaps further illuminate how the organization worked.

One day Mr. Ford came to me and said, "I see you've got a new assistant over in Pratt and Whitney."

"Who's that?" I said.

"Fellow named Grant," Mr. Ford said. "A crewman Henry brought out from Yale. He's got a better office than you have."

This was the first I knew about Grant. Though top personnel direction was supposed to be in my hands, Henry thoughtlessly had brought Grant out and put him in charge of personnel in the P & W Building without mentioning the fact to me—though he had told Sorensen.

It annoyed me that Henry had done this without telling me. I knew it would upset older men over there, men who might have rated the job. I was further annoyed because in telling me this, Mr. Ford was chiding me for not knowing what was going on. Besides, the whole thing went against our established policy.

Mr. Ford, Sorensen, and I had agreed

that we'd build up executives out of personnel in the plant, rather than bring someone in from outside. This was always Mr. Ford's idea of how things should be done. In fact, there was a famous incident that occurred before my time, which Mr. Ford liked to tell and retell, to illustrate how he felt.

It seems that years ago an engineer named Harold C. Wills told Mr. Ford that he needed a metallurgist. They were in the plant at the time, with Sorenson.

"All right," Mr. Ford said. He turned to Sorensen and pointed to a man who was sweeping the floor. "Make one out of that fellow over there with the broom."

They made a metallurgist out of this fellow, whose name was Wandersee, and a good one, too. And ever after Mr. Ford would tell this story when someone suggested bringing in an executive from outside.

Anyway, young Grant posed a situation that I couldn't let stand. Since Mr. Ford had said that Grant was a crewman, I thought he might get a little tough if we tried to bounce him. So I sent Frank Holland, who was six feet two and husky, and Jimmy Brady, who had been a good boxer, over to see Grant, to tell him that whoever put him there had no right to do so.

These two men went over to the P & W

Building, but they didn't have any trouble. Grant was a crewman, all right—he was a coxswain, and not much more than five feet tall!

Edsel heard about the thing and came to see me, indignant that I had sent "two big thugs" to intimidate the poor fellow.

I told him that I'd heard Grant was a crewman, and had thought he was a big husky fellow who might get tough.

But Edsel wasn't amused. "Well, I talked to Father, and he didn't seem to know anything about it."

"Why, he's the one who told me about it in the first place," I said.

We discussed it further, and after a while Edsel saw the humor in the whole thing, and laughed about it.

Grant, too, turned out to be no problem. He came over to see me and said, "Now, just what can I do around here?"

I told Grant that the union had taken over our record files in the employment building, and were giving fellows unmerited raises and were giving them credit for time when they were not even in the plant. I said, "If you can steal that department, you can have it."

We didn't have a union in the Administration Building—I had an oral agreement with Phil Murray that they wouldn't organize our office. So, one or two at a time, Grant moved members of the payroll record department over to the Administration Building, until he got the whole department over there, and we again had control over our records.

I made Grant the head of this department, and he worked out just fine.

Chapter 20

EVERYTHING seemed to conspire to create differences between Edsel and me.

For instance, this kind of incident:

During the early part of the war, Edsel was president of a German-American shipping line. Of course, that was fine material for a scandal if it had leaked out, and Mr. Ford asked me to go up and ask Edsel to resign from the shipping company. It was an unpleasant task, and no concern of mine, but I couldn't refuse.

Edsel agreed. He showed me his letter of resignation, saying, "Well, there goes a million bucks." The implication was plain: I personally was depriving him of a million dollars.

And then there was the time Edsel and I

got into a conflict about how to handle a union dispute.

We had got into a situation with the union where we were pretty much at loggerheads over a problem, and Edsel asked me to see a certain Army man along with a union delegation. This fellow was a former insurance salesman whom the Army had made into one of their labor representatives. Apparently he had got to Edsel alone some way at Grosse Pointe, and injected himself into the situation.

I told Edsel that I would see the union any time, but that I wouldn't see the Army representative because he had no background in labor relations, and I didn't think he knew what he was doing.

"If you won't see them," Edsel said, "I will."

"Then you'll be seeing a whole lot of them," I told him.

Edsel conferred with the Army man and the union delegation. And sure enough, the next week, when they had a new problem, they wanted to see him again. As I had foreseen, the union figured they had got the ear of someone "higher up," and that it would be more advantageous for them to deal with Edsel.

However, one meeting was enough for Edsel, and he referred them to me. Some of the union leaders promptly decided they were getting the run-around. A small group made up their minds to force their way into Edsel's office. They pulled the wires off some trolleys and organized all the men aboard into a delegation. Then they marched into the Administration Building, clambering over some of the fine furniture up on mahogany row en route. Edsel was unwilling to see them, and slipped out a back door. Finding their quarry gone, the men stood on Edsel's desk with their hob-nailed boots and made speeches.

Edsel came down to me and said, "This is your baby. I'm going to a bank meeting." And off he went.

That was Edsel's last effort at being democratic with the union. And I guess the whole thing didn't do anything to make either of us feel any more friendly.

There were other situations when I became a kind of mediator between Edsel and his father, and I don't know how much that helped our relationship, either.

For example, back in May of 1939 the Department of Justice brought an action charging the Ford Motor Company, along with General Motors and Chrysler, with violating the Sherman Anti-trust Act. The charge was that these three corporations, through their separate finance companies, were trying to monopolize the financing of their automobile sales. A federal grand jury returned an indictment.

The result was plenty of excitement in the Ford Motor Company.

The Universal Credit Corporation, the Ford financing agency, had been established in 1928 with Ernest Kanzler as its president. Mr. Ford had put Kanzler there to appease Edsel, whose resentment at Kanzler's departure had almost been strong enough to cause him to leave the company. When Mr. Ford put Kanzler at the head of the UCC, he said that he thought that was the right place for him, and added, "Kanzler and Edsel both ought to be bankers."

When the indictment was brought against the Ford Motor Company, Edsel was deeply upset and fearful. At Edsel's request, I hired Tom Lamphere and Eddy Clemment to go to Indiana, where the legal action was taking place. Their job was to learn what we had to do to settle the matter. They found out that all we had to do was to sign a consent decree, agreeing to stop the alleged practices.

Edsel was all for this, and so was Longely, who was representing Edsel. But Mr. Ford had a completely different reaction. He was mad. He wanted to fight. He said he wouldn't sign a consent decree, and wouldn't let anyone else do so.

This stand, of course, put both Longely and Edsel behind the eight ball.

Edsel asked me to intervene with his father. As I said earlier, Edsel could have gone to his father himself, but didn't because he would never argue with Mr. Ford. He knew that I would.

Edsel said, "Sure, Father wants to fight. But I'm the president of the Ford Motor Company, and I'm the one who will go to jail." Tears came into his eyes.

I talked to Mr. Ford about the matter. I told him how upset Edsel was, and that affected Mr. Ford. I also told Mr. Ford that it was General Motors' hope we would fight the case, and gave him proof of that. That softened him up right away.

Finally Mr. Ford agreed to let us settle. We sent word to the district attorney's office that we'd sign a consent decree, and that ended what was a tense and difficult situation. But, understandably enough, it did nothing to ease the situation between Edsel and me. No man enjoys the necessity, real or fancied, of using an intermediary to plead with his own father.

Our dealings with John Bugas did nothing to bring Edsel and me together, either. A number of events conspired to bring Bugas, then head of the Detroit FBI office, into contact with the Ford Motor Company. One of these was a series of serious thefts of tires and parts at the plant. When the thefts were first observed, we went to work on it, and rapidly gathered all the necessary information. Then, at Edsel's request, John Bugas stepped into the case.

He came to see me about it, and I gave him all the information in our files.

I then went to California for a vacation. While I was there, Bugas broke the case for the newspapers. They gave the story a big play, but were careful to call me up first out in California, and assure me they were going to say the thefts had been uncovered "with the assistance of Harry Bennett."

Sometime after this, around the beginning of the war, while Bugas was still with the FBI, Edsel began using him as his own source of information. Shortly, Bugas began calling in a number of my employees and cross-examined them about what was going on in the company. Of course, these men told me all about it.

I asked Capizzi to talk to Bugas, and tell him that if he wanted to know anything about me, he should come to see me, and I'd be glad to tell him whatever he wanted to know. If there was one thing I did not appreciate, it was being investigated behind my back.

One day when Edsel was in my office I took the matter up with him. I told him that I was sick and tired of having my men called in by Bugas.

"Well," he said, "we've got to do something. They're carrying the plant away."

By "they" he meant his father. Edsel

was worried about the money Mr. Ford was spending.

"Why don't you tell your father how you feel?" I asked.

"Why don't you?" Edsel said.

So I told Mr. Ford. But he had no idea of letting someone else tell him what to do with his money. He said, "If Edsel doesn't keep his nose out of what I'm spending at Greenfield Village and around, there won't be enough left in that plant to build a Chic Sales."

All in all, Edsel and I were having it hot and heavy by around 1942. Looking back, I think that most of my hostility arose out of Edsel's general attitude toward me. Edsel was a very democratic fellow in his own set, friendly and well liked by those in his circle. But he hadn't the respect I thought he should have for men who came up through the ranks, the hard way. He was always belittling me, and I think that, more than anything else, got under my skin.

Sorensen realized how unhealthy for the company this situation was, and he tried to reason with Edsel. I don't know all the arguments Sorensen used to try to persuade Edsel to bury the hatchet, but among other things he said, "After all, I work with the S.O.B., and I can't bear him any more than you can."

Edsel reported this statement to Mr.

Ford, perhaps hoping that his father would be influenced against me. But Mr. Ford came right over and told me, adding, "Sorensen's no friend of yours, Harry."

I never believed for a moment that Sorensen had said that seriously. I understood what he was trying to do, and how he might resort to such a statement in an effort to influence Edsel to take up the peace pipe. With all respect for Mr. Ford, I have never needed him or anyone else to tell me who my friends are.

As the reader will recall, when America got into the war the country's sources of raw natural rubber were all but shut off. We had our own tire plant at the Ford Motor Company, and began making some poor attempts at manufacturing synthetic rubber. The experiments were not too successful.

With his avidness for information and his obsession on the Du Ponts, Mr. Ford got to wondering why the Du Ponts weren't making synthetic rubber, and didn't seem to have any intention of doing so. He asked me to find out for him. Perhaps he thought he'd get a clue to the solution of our difficulties, or perhaps he just wanted to know what his "enemies" were up to.

Anyway, someone found the right investigator for me, a man who had been a

Michigan state legislator. I sent him down to Wilmington.

The man picked up some information in a few days, and started back home. But in Philadelphia he dropped dead—from natural causes, I must add.

So I had to start out all over again. I found a second suitable investigator, told him how far the first man had got, and sent him to Wilmington. He worked for about two weeks, reporting some information, and incidentally stumbling across what turned out to be the atomic-bomb project.

Then, by strange coincidence, the second investigator had a heart attack and dropped dead.

When this second death occurred, Mr. Ford came to see me, all excited. His worst suspicions in regard to the Du Ponts had been confirmed.

"Harry, don't send any more men down there," he said. "By God, they're killing them!"

That ended Mr. Ford's interest in why the Du Ponts weren't making synthetic rubber.

Mr. Ford had been saying, ever since we had got into the war, that perhaps we ought to do something to help the Russians. Now Edsel and Sorensen told him that our government wanted to buy our rubber plant for the Russians and ship it over there. They suggested that we sell it, and Mr. Ford, after making a few comments on the government, agreed.

So the rubber plant was dismantled and the government shipped it to Russia, lock, stock, and barrel.

That seemed to be that. But about two months later Mr. Ford came to see me, all excited, with an idea on how we could improve on our synthetic rubber.

"Why, Mr. Ford," I said, "that plant went out of here two months ago."

"What!" he exclaimed. "Where did it go?"

"They sent it to Russia," I said.

"Why did they do that?"

"Because you said they could."

"I did not!" Mr. Ford said. "I never heard a thing about it!"

We argued back and forth, until finally he said, "Well, I don't care if I did say they could. They certainly worked awful fast!"

In time, Mr. Ford forgot about synthetic rubber. But not the Du Ponts.

In the spring of 1942 Mr. Ford decided that he wanted to needle the Du Ponts. He told me to get word to them that he was going to build a huge new laboratory on his farm. Of course, he had no such intention. How he got the idea in his head I don't know, but he thought that such a "tip" would get his archenemies worried and wondering just what he was up to.

Silly as the whole thing seemed to me, I did as he asked.

Now, as it happens, the government began to get worked up over the question of housing for Willow Run employees. They wanted to build temporary housing near the bomber plant. Most of the land around the plant belonged to Mr. Ford, and was a part of his farms. So the Federal Public Housing Administration began surveying some land of Mr. Ford's, setting out about seven hundred surveyors' stakes. Mr. Ford looked at them as we drove by, and then did a double take.

Mr. Ford blew his top. He was convinced that the whole thing was a plot, and that the Du Ponts were behind it all. It was perfectly clear to him that the Du Ponts had got the government to build there so that he couldn't put up the laboratory—which, of course, he had never intended to do in the first place.

"You just turn around now," Mr. Ford said. "We're going to pull those stakes out."

So I had to turn the car around and go back. Mr. Ford and I got out, and I had to help him pull out I don't know how many of those stakes. When Mr. Ford was tired we went back to the plant, and then I had to send some men out to finish the job.

There was no way to budge Mr. Ford from his idea, and he had us fighting the government's housing plans tooth and nail. As it happens, there were some good grounds for taking this position. We foresaw that a temporary housing settlement would become a severe community problem, and it did. We also felt that the government wanted to build more dwellings than were necessary, and that too proved to be the truth.

By fall, however, Mr. Ford cooled off. In October he sold the government 295 acres of his Willow Run property for a housing project.

Chapter 21

As THIS country came into the war, one of the most distasteful duties I had was the performance of certain tasks for the three Ford grandsons. I thought they should be treated like everyone else in the service, but Mr. Ford was very much concerned for their welfare.

Henry served in the Navy—at the Training Station at the Rouge, and at Great Lakes Training Station. Benson entered the Air Force, where he became an aide to General "Sudden Sam" Connell. William, the youngest, was in the Naval Air Corps training unit at the University of Michigan. Mr. Ford became so concerned for them, worrying and fretting, that he asked me to do more things for them than I can recount.

This got to be a burden and an annoyance to me. At one time Mr. Ford kept telling me, over and over, to be sure that Henry had everything he wanted, particularly transportation. I did as Mr. Ford asked.

Henry resented this, and accused me of keeping track of his movements. Apparently he communicated this idea to his father because Edsel wired me from Miami, asking me to please let Henry go on his own.

I thought of all I'd been put through for those boys, and I sent a return wire to Edsel telling him I thoroughly agreed—Henry *should* be on his own.

I told Mr. Ford about the telegrams. I was angry and thoroughly disgusted, and I told him I wanted to quit and go into the service myself.

"Now," Mr. Ford said, "don't you do another damn thing for him."

Mr. Ford's attitude didn't console me much, but at least it released me from any further obligation toward Henry's welfare.

At about this time, also, there began a chain of events that was to have a profound effect on the Ford Motor Company, though we could not have foreseen this at the time.

Mr. Ford didn't like either Liebold's or

Cameron's sympathies in the war (Cameron was violently pro-British), and he began baiting both men. For quite a while this was Mr. Ford's unvarying morning routine: First he walked from his office (in the Dearborn laboratories) into Cameron's office. He stayed there a short time, getting Cameron riled up about the war. He next went farther down the hall to Liebold's office, and got Liebold boiling on the same subject. Then he came over to the Rouge to see me, to tell me with delight how he had left the two men quarreling.

This growing antipathy toward Cameron on Mr. Ford's part helped bring to a close Cameron's radio career.

For some time I had been trying to convince Mr. Ford that the Ford Sunday Evening Hour, a program of classical and semi-classical music, was a mistake. I pointed out to him that we were selling low-cost cars, and that this type of program did not appeal to the majority of people who made up our market.

In this opinion I was sincere. But in addition to wanting more for our advertising dollar, I hoped to get Cameron off the air as commentator, because I felt he was doing the company no good.

After many long discussions, I got Mr. Ford to think of putting Paul Whiteman on the air instead of a symphony orchestra.

Mr. Ford went to Cameron and told him he was thinking of making this change.

Always the consummate actor, Cameron worked real tears into his eyes. "I never thought I'd see the day when the Ford Motor Company would sponsor the King of Jazz," he mourned.

Of course, that was a shrewd thing for Cameron to say. Mr. Ford's distaste for jazz was well known. Mr. Ford promptly forgot about Whiteman's ability to play other types of music, and backed out on the whole deal.

However, finally, in 1942, we did get Cameron off the air, in a roundabout manner. By this time, Mr. Ford was all for it.

With Cameron's prestige on the decline, then, an important change took place in our handling of public relations.

One day Edsel came to see me and asked what I thought of bringing Steve Hannagan, the well-known publicist, into the company to handle our public relations. Edsel knew my dislike for Cameron, and he didn't like the man much himself. He said he thought it might be a good idea to bring in Hannagan and make him a kind of co-ordinator of public relations, clearing all news through him.

Next Sorensen came to see me. He said he thought Hannagan was a good man, and why didn't we bring him in? I said it was all right with me, and began wondering why the whole thing was being dumped in my lap.

Apparently both of them had been talking to Mr. Ford, because soon he called me and asked me to find out what I could about Steve Hannagan. I told Mr. Ford that as far as I knew, he was the best in the business.

So, in June of 1942, Steve Hannagan was brought in. And on that fact hang many of the historic events that followed in the Ford Motor Company.

Production at Willow Run began about May 1942. But for over a year, production was a mere trickle. While the publicity department kept on putting out more and more grandiose promises to the public, Willow Run labored mightily to produce a mouse. By 1943 the plant was the subject of widespread gossip, and early that year it was made the subject of an investigation by both the Office of War Information and the Truman Committee. The latter, in the summer of 1943, published a highly critical report.

What was wrong?

There were a number of things wrong. Mead L. Bricker, now vice-president without portfolio of the Ford Motor Company, was in charge of the Willow Run bomber plant, under Sorensen's general supervision. Though Sorensen had put him in that job, Bricker, in my opinion, wasn't backing up Sorensen. He accused Sorensen, to Mr. Ford, of ruining morale at the bomber plant. But meanwhile Bricker was busy trying to find out what people thought of Bricker; and it was that, in my opinion, which raised hell with morale out there.

When Hannagan came to the company, Bricker saw in him the instrument he wanted.

Mr. Ford suggested to Hannagan, when he came in, that he build up some of the other executives around the place, particularly Sorensen. Hannagan went to work on that, and he really did a job, too. Magazine articles, news items of all kinds began appearing in the press, portraying Sorensen as the production genius of the Ford Motor Company.

Mr. Ford and Sorensen had gone along for years in a happy, close relationship of mutual confidence. But now the suggestion was made to Mr. Ford that Sorensen was trying to get all the publicity for himself, and was trying to put Mr. Ford in the shade. And though Mr. Ford had himself asked Hannagan to build up Sorensen, suspicions began to seep in.

At first, though, the effects were not noticeable, and Bricker only managed to draw unfavorable attention to himself. I suppose it seemed like a good idea to him to report to Mr. Ford out at the bomber plant that Sorensen was sending out letters signed "General Manager."

"I'm the general manager here," Mr. Ford said, "and if I ever find out there's another general manager around here, I'll fire him."

Mr. Ford and I left the plant shortly. He told me what Bricker had said, and indicated he didn't like Bricker.

As we drove along back to Dearborn, Mr. Ford was silent for a time. Though I did not guess it, his mind was busy working out a trap for Bricker.

Suddenly he said to me, "Harry, I want you to make Bricker general manager of the bomber plant. Go tell Edsel to do that."

"All right, Mr. Ford," I said, "if that's what you want."

It took a little while before I realized Mr. Ford's scheme. His intention was to have someone else make Bricker "general manager"—and then, as soon as Bricker signed a letter that way, fire him for it.

The next day I went to see Edsel for the purpose of relaying Mr. Ford's request, but before I could bring up the subject Mr. Ford himself walked in.

"Edsel," he said, "I want you to make Bricker the boss out there."

"What do you mean, the boss?" Edsel said.

"Well, you know," Mr. Ford said. "Make him the head cheese."

He was, of course, being foxy, and avoiding use of the phrase "general manager," so that later he could deny ever having given such an order. (He had used that phrase with me, because he was confident that later on I would never tell anyone.)

Mr. Ford went out then, and Edsel said to me, "I see through this, all right, but I'll make him general manager."

Bricker was duly notified of his "advancement" out at the bomber plant.

As soon as Bricker signed his name to a letter with his new title, Mr. Ford was ready. He had *told* him, Mr. Ford said, that if he ever found another general manager around the plant, he'd fire him.

That should have been the end of Bricker. But Sorensen, unaware of Bricker's activities, stepped in and stopped the firing.

However, Mr. Ford didn't give up that easily, and neither did Bricker. A short while later, when we were again out at the bomber plant, Bricker resumed his offensive against Sorensen. Every time Sorensen came out there, he said, he upset things.

"Well," Mr. Ford said, "the next time Charley comes around here, you just tell him to get the hell out."

Again, as we drove away from the bomber plant, Mr. Ford was silent a while as we rolled along. Then, suddenly, he burst out laughing.

"God, isn't that going to be funny when Bricker tells Charley to get out of the plant?" Mr. Ford said. "Charley will fire him on the spot."

Somewhere, Mr. Ford's scheme backfired. But it was a good try.

Meanwhile, production at the bomber plant still wasn't satisfactory, and everyone's nerves were stretched to the breaking point. And as a final complication, we were soon faced with a situation of extreme racial tension in the Detroit area. We got a tip that if race tensions broke into violence, as seemed likely, it was going to begin at the Rouge or at Willow Run, since nearby Ypsilanti had a large Negro population.

In this general and almost unbearable tension, there occurred the only quarrel that ever took place between Sorensen and myself.

The whole situation in regard to Negro employees had become something of a mess. In the hiring of Negroes, as in the hiring of other men, we were trying to pick the best of those who applied. In general, we wanted to get the healthiest and best educated men we could. As a result, many we considered unfit were turned away. The government began putting heavy pressure on us to hire more Negroes.

Then a situation came to light that shows something every executive knows: how close you can be to something, and yet have things get past you.

I was discussing the whole thing with Mr. Ford when he said, "Why, you're hiring them down South."

"No, we're not," I said.

"Yes, you are," he insisted.

So I investigated the matter, and found out that sure enough, Miller, our employment man, was down South hiring Negroes as fast as he could.

What had happened was that Bricker and others had become panicky at the thought of not being able to meet their labor requirements, and had asked Miller to do this. But by the time the men Miller had hired came up, they weren't needed.

Well, I brought Miller right home again.

In an effort to work out the whole problem, I had a conference with Roscoe Smith, superintendent of the bomber plant under Bricker. We discussed the thing from all angles, and finally agreed that we already had the best Negroes we could get in the bomber plant. We further agreed that the best thing to do, in view of that fact, was to make our Ypsilanti plant all-Negro, and in that way absorb the others.

Shortly I met with Sorensen, Bricker, and Smith at the bomber plant, believing that this plan worked out by Smith and myself was all set. But to my astonishment, it suddenly appeared that such was not the case.

Sorensen said, "We just don't want any more Negroes. We've got more than our quota now."

"Why, we're all agreed on this thing," I protested.

Sorensen turned to Smith and said, "Do you think this is a good idea?"

"Why, no," Smith said, shaking his head.

Seeing the whole thing blow up like that, when Smith had just agreed with me previously, was too much. Smith was a big man, about six feet four, but the tension that had been building snapped in me, and I swung at Smith over Sorensen's shoulder.

Smith went down and stayed there. Furious, I turned on Sorensen. I guess I would have hit him too, but he's the kind who won't fight with anyone. He just got up from his chair and walked out of the room.

Only much later did I recall that Bricker had stayed in the background throughout, smiling.

I went home and told my wife what had happened. A little later Bricker called me up. He said Sorensen had come back to the plant looking for me, and quoted Sorensen as saying, "Why, I can lick six like him."

That didn't sound like Sorensen to me,

but I was still worked up, and all ready to go over there and find Sorensen. But my wife stopped me.

"I can smell a rat here," she said, and began trying to reach Sorensen by telephone.

She couldn't find him as it turned out. But by the time she'd finished trying I had cooled off, and I began to realize what had been going on.

A couple of days later my wife and I met Sorensen at a party. He was very friendly to me, and neither of us spoke of the incident either then or later. Evidently Sorensen knew the score better than I thought he did.

After this, I got in a woman named Josephine Goman to handle our housing problems and the distribution of Negro help. She worked at this with the utmost intelligence, tact, and fairness to all, and handled the thing in a way that deserved a medal.

Shortly after I had hired Josephine Goman, Mrs. Ford called me up and complained about her; she'd heard things she didn't like, she said.

I told Mrs. Ford that Josephine Goman was doing a wonderful job, and that we needed her badly. I suggested that she discuss the matter with Mr. Ford. I told Mr. Ford about the call, and happily he backed me up.

After that, our problems with race tensions were never again so severe.

The production problem, though, had still not been licked.

By this time Slim Lindbergh had come to work for the company at the bomber plant, as a consultant. He made a trip to Washington and returned to tell us that he had heard some very discouraging things about Willow Run. Intimating that the government might take the plant over itself, he said, "You'd better be prepared, and see if you can't do something about it."

Mr. Ford had been getting more and more upset by the way things were going out there, and now he sat down with Sorensen and me. He said, "What are we going to do about this place?"

We discussed the work out there, and I said I thought we could get things rolling if we set up a new system of station inspection. I suggested sending out Ray Rausch, Al Kroll (who was chief inspector at the Rouge), Bill Comment, and Harry Mack. Sorensen promptly agreed, and said he would release all four of them from their work at the Rouge.

When told of this plan, Bricker disapproved of Kroll and Comment, but it didn't take Sorensen long to convince him that they were needed badly.

So a new system of station inspection

was set up at Willow Run, and immediately after that things took a new turn. It was uphill all the time after that.

By 1944 the Willow Run bomber plant was making one of the important industrial contributions to the war effort.

Chapter 22

Although I haven't spoken of it for some time, Mr. Ford's feud with Edsel's Grosse Pointe friends and relatives was by no means ended. His feeling toward anyone he thought was influencing Edsel was a thread that ran through all his days.

Now Mr. Ford sent John Gillespie into the Purchasing Department, telling him to check on any accounts held by Grosse Pointers. Mr. Ford wanted to ferret out any who might be making money on the Ford Motor Company, and freeze them out. He told Gillespie, "You can have any accounts held by any of these people who used influence with Edsel to get them."

Gillespie raised hell in Purchasing, since he carried in there his philosophy of creating confusion and getting people fighting among themselves so that he could run off with the plums. I got repeated complaints about Gillespie from A. M. Wibel, who was a vice-president and head of Purchasing, and from Edsel.

Consequently, I tried to keep Gillespie out of the building.

As soon as I did this, I received a call from Mr. Ford, asking me to meet him at the Dearborn lab the next morning. I did so, and he asked me why I didn't want Gillespie in the plant. I told him it was because of Wibel's and Edsel's complaints.

"Well," Mr. Ford said, "I know why they don't want him up there—he's giving me too much good information. You just tell John to go right ahead. He can have any account that was cooked up in Grosse Pointe."

I thus lost all control over Gillespie, and he did pretty much as he pleased, coming into the plant whenever he felt like it.

Wibel was so furious at John's carte blanche that he went to Wright Field (all our work was military by now) and complained about Gillespie. In his rage, Wibel made an attempt to link me up with Gillespie's activities. The accusation was so absurd that even Edsel was angered.

I was myself so indignant that I told Mr. Ford I intended to resign. However, Mr. Ford wouldn't hear of it. He was already furious with Wibel, both for his accusations against me and because he had taken this matter outside the company. He ordered me to fire Wibel.

I refused to do this. Though Edsel was angry at Wibel, still Wibel was Edsel's friend, and I didn't want to be involved in what was at best an awkward situation. So Mr. Ford told someone else to speak to him, and in April of 1943 Wibel left the company.

In that year, 1943, Mr. Ford's health began going downhill. His physical and mental decline was precipitated by an event that had the Ford family in an uproar. It's rather a long story, but an important one.

Some years earlier, Mr. Ford caught on to the fact that people used his well-publicized prejudices to get things out of him. Whether it was a reporter looking for a hot interview or a politician looking for a campaign donation, nearly everyone commonly began Jew-baiting as soon as he got in to see Mr. Ford. Even so eminent a man as Wendell Willkie tried this tactic when he wanted to get a half-million dollars from Mr. Ford for his campaign for President in 1940.

But, as I said, Mr. Ford got wise to this. Usually the person trying to get something out of him would begin railing at the

"Jews around Roosevelt," often naming people who weren't Jews at all. To the astonishment of the other person, Mr. Ford would put a stop to the whole thing with an unexpected response.

"Well, that's all right," Mr. Ford would say. "If 'they' weren't around him to handle the money, this country would go broke."

Though Mr. Ford said this largely for effect, he really meant it. He highly approved of Morgenthau, for example, and greatly admired Baruch.

However, Mr. Ford was just as often gullible.

He had certain influential friends who made speeches at Jewish functions, and in general took a neighborly public attitude. But when they were with Mr. Ford, they egged him on in his prejudice, or else dished out some of their own dirt. I've heard them say, "That's right, Mr. Ford, you just go ahead and give it to them."

Well, one day in 1943 I pointed out to Mr. Ford that these fellows were playing both ends against the middle. They were always ready to stick Mr. Ford's neck out.

Mr. Ford realized that this was so, and he was deeply indignant at his friends' behavior, once he saw it in its true light. He decided that he wanted to get out a statement to the papers exposing these people, and I encouraged him in it.

Mr. Ford got up a red-hot statement, which ended with the phrase: "Cows, horses, pigs, and creeds will someday disappear from the earth."

He gave this statement to John Thompson, one of Steve Hannagan's men, and told him to get it to the press.

Then the Ford family blew up. Ann Ford, young Henry's wife, said, "There'll be a scandal in the Ford family if that's ever printed." Others pitched in. One minute Mr. Ford told Thompson to go ahead and send it out or he would be fired, and the next minute the family told Thompson not to dare send it out—or he would be fired. This battle raged until, finally, Thompson left the company. Mr. Ford angrily stumped off to Georgia, still determined to publish that statement.

Some members of the family, still bent on stopping him, followed Mr. Ford to Georgia, and were sent packing right home again.

In the end, the statement was given to the press, but in such a mutilated form that it meant nothing, and was regarded by the public as just one more eccentric statement from Mr. Ford.

Mr. Ford never forgave the family, and in his anger he went on his own version of a hunger strike. Always a food faddist, he now took on eating habits that could hardly fail to break his health.

For one thing, Mr. Ford became convinced that sugar was not a food, but a danger to the human body. He wanted to "prove" this to everyone. He was like that; he couldn't tolerate disagreement. If you believed differently than he did, he'd keep nagging at you and nagging until you either changed your mind or said you did.

To "convince" Hub McCarroll, chief chemist of the company, Mr. Ford bought a huge magnifying glass and showed crystals of sugar under it to Hub. Of course, the crystals looked sharp and jagged, and Mr. Ford's contention was that these crystals acted like knives on human tissue.

"Yes, Mr. Ford, but look at this," McCarroll said, and put a drop of water on the crystals. They dissolved at once, of course. This infuriated Mr. Ford, and he wanted to fire McCarroll.

In addition to this, Mr. Ford took up a diet consisting solely of cracked wheat. "You can live as long as you want," he used to tell me, "as long as you only eat cracked wheat."

But Mr. Ford's declining health was not the only tragedy in the Ford family. Edsel was mortally ill with cancer and was rapidly approaching death.

Mr. Ford was always distrustful of the medical profession. At one time he wanted to kick all the doctors out of Henry Ford Hospital and install chiropractors instead.

At first he refused to believe that Edsel was seriously ill, and kept this up until Edsel underwent an operation in 1942. Then, though he was forced to see how ill Edsel was, he refused to believe that Edsel had cancer. Up to the very end he insisted that Edsel had undulant fever. So strong was this conviction that he stopped the use of milk in Greenfield Village, and wanted Dahlinger to get rid of all the cows on his farms. (That's the reason he included cows in the press statement I just told about.)

In addition to this "undulant fever" theory, Mr. Ford blamed Edsel's ill health on the "ear-piddlers" around him, and Edsel's "high living." In his zeal, Mr. Ford checked on all parties that Edsel gave or attended.

About a year before Edsel died, this practice had some dramatic results.

Kanzler arranged a big birthday cocktail party for Edsel in Washington. Mr. Ford was, of course, told nothing about it. I don't remember the whole guest list, but there were a lot of people invited—Donald Nelson, J. Edgar Hoover, Jim Farley, and Steve Hannagan, among others.

Cameron heard about the party and informed Mr. Ford. He hit the ceiling. "They're trying to kill Edsel!" he said, and asked me to check on the story. I called our Washington sales manager, and he verified the whole thing. Mr. Ford was furious.

It wasn't only that some drinks had been pushed toward Edsel. The guests had occupied themselves in a rather unusual way—they had planned a big reorganization of the Ford Motor Company! I don't recall all the details, but Donald Nelson was to be production manager, Jim Farley sales manager, and so on.

Well, naturally these plans didn't please Mr. Ford much, and he raised the devil about it. But he also managed to make the whole affair useful to him.

By now, Mr. Ford's jealousy for Sorensen's new public prominence was working on him. Its immediate effect, at this time, was to make Mr. Ford furious with Hannagan. He wanted to get Hannagan out, but as usual he didn't want to do it himself.

So Mr. Ford told Sorensen all about the Washington party and the reorganization they had planned. Actually, in their scheme they had allotted some place or other to Sorensen, but Mr. Ford concealed this fact. He said, "Well, Charley, I notice they didn't have any room for you in their setup."

This made Sorensen angry, and as Mr. Ford had expected, his wrath turned on Hannagan, who had been there. Hannagan left the company.

Everything that could have been done for Edsel by medical science was done. But it was not enough, and he died in May of 1943.

Edsel's death was a blow to Mr. Ford—the greatest single catastrophe he ever suffered. At first he fell back on his theory of reincarnation, and said to me, "Well, Harry, you know my belief—Edsel isn't dead."

But he couldn't deceive his own emotions very long, and soon he told me, "I never believed anything like this could happen to me."

Mr. Ford asked me if I intended going to Edsel's funeral, and I told him, "No, I couldn't be that hypocritical." I knew that Edsel had despised me, and felt that the honest thing to do was to stay away. Afterward, Mr. Ford came to tell me about the funeral, and when I heard about the phony tears shed there by some of Edsel's enemies, I was glad I hadn't gone.

Even now, with Edsel dead, however, Mr. Ford still went on insisting that Edsel had had undulant fever; he caused notices to this effect to be posted around the plant when Edsel died. At the same time, his hostility to the medical profession became more bitter, and he blamed Edsel's death on the doctors. A short time after Edsel's death, in the presence of others I told Mr. Ford that he should give a million dollars to the Cancer Fund. He exploded at this, and later, when we were alone, raged at me for having made the suggestion.

Well, Edsel and I had this much in common: When he went, he left a lot of "orphans" in the plant; and when I left, so did I.

With Edsel gone, there was the question of who was to succeed him as president of the company. Mr. Ford talked this over with me. Henry, Edsel's eldest son, was only in his late twenties, and had had no real connection previously with the business. So I suggested to Mr. Ford that he make Sorensen president, and let him hold that place until Henry was mature enough and capable enough to take over.

"You're absolutely right, Harry," Mr. Ford said.

So in Mr. Ford's mind, Sorensen was president of the Ford Motor Company—for exactly eight hours. And I think that when he reads this, it will be the first he ever knew about it.

As soon as Mr. Ford announced his intentions to the family, Mr. Ford told me, certain members of the family objected violently. Campsall lit into me for having made the suggestion, saying that I was "losing my mind."

I guess it got pretty bad; I don't know all of what went on. But eight hours after having decided to make Sorensen president, Mr. Ford called me up. He said, "I'll settle this—I'll be president myself. I don't think anyone will have anything to say about that."

So Mr. Ford took up again the title that he had handed to Edsel, without ever passing on the power that went with the scepter, and he kept the title up until his final illness.

Within a matter of a few weeks, Henry was released from active Navy duty. He was made "executive vice-president" of the company, and took up offices in the Administration Building in the Rouge.

At the same time—in June, to be exact—Mr. Ford made me a director of the Company. This may sound very impressive, but only until the reader understands the nature of directors' meetings at the Ford Motor Company.

Directors' meetings were the funniest thing in the world. If a movie is ever made about Mr. Ford, they would furnish the comedy.

These meetings had no purpose other than to comply with the law. When Mr. Ford failed to show up, it was pretty funny, because all the directors dared do was conduct cut-and-dried business—putting their stamp of approval on what Mr. Ford had already done, or on what they knew he'd approve. None of them ever dared undertake any action that involved initiative, with the sole exception of Sorensen, who sometimes did this, to the great discomfort of Craig and the others.

One of the high moments I recall was when the Board of Directors solemnly voted a loan to Mrs. Edsel Ford to pay her inheritance tax, after she had resigned as a director in order to make such a loan legal. I'm sure I must have seconded that one.

However, the directors' meetings reached their peak of humor on those occasions when Mr. Ford did show up.

Mr. Ford would come in, walk around, shake hands with everyone, and then say, "Come on, Harry, let's get the hell out of here. We'll probably change everything they do, anyway."

Chapter 23

EDSEL'S death worked a profound change in Mr. Ford. It seems impossible to describe how deeply the loss of his son hit him. After that, he wasn't anti-Semitic or anti-Catholic or anything else. He was just a tired old man who wanted to live in peace. He reached a point where he didn't want to see either Liebold or Cameron. He was ready to do almost anything he thought Edsel might have liked.

Mr. Ford was even ready to make peace with Kanzler. I tried to arrange for Kanzler to come out to lunch, and Mr. Ford was willing to sit down with him. I asked Henry to extend the invitation to Kanzler, but he didn't come. Perhaps Henry failed to invite him, or perhaps he simply didn't want to.

After Edsel's death Mr. Ford was disturbed about the relationship he had had with his son. He couldn't keep away from the subject. He'd bring it up with me, and we'd discuss it a while, and then he'd say, "Now, we aren't going to talk about it any more." But then he'd come back to it, again and again.

"Harry," he once said to me, "do you honestly think I was ever cruel to Edsel?"

It wasn't easy to answer that one directly, and I temporized: "Well, if that had been me you'd treated that way, it wouldn't have been cruelty."

But Mr. Ford wasn't satisfied. "Why don't you give me an honest answer?"

So I said, "Well, cruel, no; but unfair, yes." And then I added, "If that had been me, I'd have got mad."

Mr. Ford seized on that. "That's what I wanted him to do—get mad."

"Well," I said, "you sure made him mad at me."

And that's the way it went.

While Mr. Ford's attitude toward the world changed after Edsel's death, his attitude toward himself never changed. He had always held himself above the rules of the game by which other people played; and that was something too deep and strong ever to leave him. And there's a story about that, too.

One day when I was driving Mr. Ford back from the bomber plant after Edsel's death, he told me something Bricker had said about me.

I got boiling mad. I said, "When I get you home, I'm going back and sock him."

Mr. Ford, in his new, benign mood, said, "Oh, well, Harry—live and let live."

Still so angry that I spoke without thinking, I snapped back, "It's kind of late in the day to say that, isn't it?"

The minute it was out, I could have bitten my tongue.

"I'm sorry I said that, Mr. Ford," I apologized. "I was sore, and just wasn't thinking."

He looked at me in surprise. "Why are you sorry?"

"I really didn't mean it," I protested.

"Oh," Mr. Ford said, "that's all right. I was thinking of people in general."

That anecdote, probably, will throw more light on what made Mr. Ford tick than any other I have told.

Meanwhile, there was something else in Mr. Ford that didn't change: his feelings about the Du Ponts.

One night when Mr. Ford called me up at my home at nine-thirty, as he did for so many years, he was tremendously excited. He said, "I'll be out and pick you up first thing in the morning. I've got some startling news. I'll show you I'm right about those Du Ponts!"

He came out to my house and got me in the morning, and told me his "startling news." It was this: Henry was getting friendly with a young man who was a new arrival at Grosse Pointe—a member of the Du Pont family! Mr. Ford said of the young man, "He was planted out there by the Du Ponts."

I saw how silly the whole thing was, and tried to minimize it. I said, "I never heard Henry mention a Du Pont."

Mr. Ford was not to be calmed, however, and finally I promised to talk to Henry about the matter.

I brought the subject up with Henry at lunch that day. He readily admitted the friendship, and said that he certainly intended to continue it.

I then talked with Mr. Ford. I told him that the friendship between Henry and young Du Pont was perfectly harmless, and that there was absolutely nothing to be alarmed about. I let Mr. Ford know that, as far as I was concerned, that ended the matter.

But it wasn't ended for Mr. Ford. He felt I wasn't taking the thing seriously enough, and was unwilling to let it drop. He sent people to see me, to try to convince me that I was wrong. Among others, he sent

Sorensen—as he always did whenever we had a disagreement. But Sorensen didn't talk to me about Henry and the Du Pont. Sorensen knew I was getting disgusted and was thinking of getting out, and tried to convince me I should stay.

By this time, things were getting so complicated, Mr. Ford's waning faculties put such a strain on me, and the place was so full of intrigue that I hardly knew what to do.

Probably a lot of the history of the Ford Motor Company will have seemed incredibly strange to the reader. But nothing, I'm sure, will be more incredible than the manner of Sorensen's leaving.

In the spring of 1943 Mr. Ford wanted to make Sorensen president of the company. In the fall Sorensen resigned.

To understand what happened, two things must be understood.

First, by the fall of that year Mr. Ford was not himself at intervals. He was losing his memory, and on occasions his mind was confused.

Secondly, the publicity build-up Hannagan had been giving Sorensen by now reached its full effect. There wasn't room in the Ford Motor Company for two "geniuses."

Early in October Mr. Ford went to his winter place at Way Station, Georgia. I went to my place in California. At the same time, Sorensen decided to take a Florida vacation. And that's when it happened.

Frank Campsall told me later that he called Sorensen in Florida, at Mrs. Ford's instigation. Campsall told Sorensen that Mrs. Ford was worried about Mr. Ford's health, and that he was fretting about Sorensen's publicity. Mrs. Ford felt, Campsall said, that it would be better if Sorensen dropped out of the picture.

The next day Sorensen announced that he had resigned from the company.

Campsall then called me up in California and told me what had been done. He told me that under no circumstances was I to mention anything about it to Mr. Ford when I got back.

I told Campsall I would make no such promise, and I said, "It sure gives me a good, secure feeling with Mr. Ford if that's the way things are handled. Why should I come back?"

That was the end of our conversation. I was now so sick of the atmosphere in which I had been living that the thought of returning to that life was almost intolerable. I went through a period of terrible inner struggle. On the one hand, I wanted to get out of it all, right then and there. On the other hand, I had been with Mr. Ford for twenty-eight years, and in a sense that was the only life I'd had. It isn't so easy to break off such ties.

Three days after Campsall's phone call, Mr. Ford called me from Georgia. He wanted to know when I was going back to the plant. I was still going through this struggle within myself, and I didn't know. I said, "When are you going back?"

Mr. Ford said, "When are you?"

"I don't know," I had to tell him.

But two weeks later I went back to the Rouge.

Meanwhile, Sorensen found that his "resignation" had not ended his difficulties with the Ford Motor Company.

Sorensen had gone to Florida in a company-owned car. When he announced his resignation, I was told, a telephone call was made from Detroit to Florida, demanding that Sorensen turn over the car to the Jacksonville branch. But by the time the call came through, Sorensen had started back home.

Some years earlier, a neighbor of mine returned from a trip to Germany and made me a present of two unusually fine clocks, of the type motivated by changing air pressure. I gave one of these clocks to Mr. Ford and the other to Sorensen.

Now, while Sorensen was driving back from Florida, someone from Dearborn went through Sorensen's office. What else was done I don't know, but Sorensen's clock was moved over to Mr. Ford's office.

After both Mr. Ford and I returned to the plant, Sorensen wrote me a letter about this clock, asking me to send it to him. I forwarded the letter to Dearborn, but got no reply.

I decided to take the matter up with Mr. Ford. He seemed confused about the whole thing, at first insisting he knew nothing about it, but then finally agreeing that he had two identical clocks in his office. "Well," he said, "I want to talk to Charley before I send it back."

I believe the clock is still in the Dearborn laboratories.

Sorensen had been the plaster that held the plant together. When word came through that he was leaving, real panic swept through the Rouge. When Sorensen left, the Rouge lost its soul.

More than that, Sorensen was Mr. Ford's real contact with production. Mr. Ford seldom went into the plant again after Sorensen left.

Soon, with Mr. Ford's memory slipping at periods, the whole thing got to be pathetic.

More times than I can remember, Mr. Ford and I would get into the car and I'd say, "Well, where will we go?"

"Let's go over and see Charley," he'd say.

"Why, he's not here any more," I'd answer.

"He's not!" Mr. Ford would exclaim. "Where is he?"

"He's gone," I'd say.

And the next day Mr. Ford would say, "Let's go over and see Charley." After a while I just didn't answer.

As for myself, Sorensen's departure and the manner of his going had a profound effect on me. I no longer had that feeling of security, so deep that I had never worried much about what salary I got. I saw that if it could happen to Sorensen, it could happen to me too. Besides that, I had always said, for more years than I can remember, that when Mr. Ford left the place, I would leave too. And I could see that he wasn't going to be around much longer.

As a result of all this, I managed to get my salary up into the $75,000-a-year bracket for the last two years I was there.

While the last two years I was with the company were pretty bad for me, I don't want to give the impression that everything was glum. There were a few bright spots. For instance, I finally got a chance to fire Liebold—the only executive I ever did fire.

In the spring of 1944 I learned some things I hadn't known about Liebold. I had a talk with Capizzi about him, and also for the first time I learned that Liebold held a power of attorney for both Mr. and Mrs. Ford.

Armed with this information, I took the matter up with Mr. Ford. I told him that with Liebold's power of attorney, he could give away just about anything Mr. Ford had.

"Oh, it isn't that bad," he said.

"Yes, it is," I insisted.

"Then you get it away from him," he said.

I then told Mr. Ford what I had just learned about Liebold. At first Mr. Ford wouldn't believe me. A couple of telephone calls, however, convinced him. He then spoke the words I had been waiting to hear for so long: "Well, you just get him out of here."

I next saw Frank Campsall and told him to take over Liebold's power of attorney, which he did. Liebold had always been paid directly by Mr. Ford, and the next step was to get him on the company payroll. Once that was accomplished, it put him within my jurisdiction, and I fired him.

Chapter 24

HENRY on occasion irritated Mr. Ford. I recall one time when Henry went East to make a public speech, and spoke out in favor of more government control over industry. In doing this, he went against everything that Mr. Ford had always stood for, and Mr. Ford was enraged.

When Henry came back I told him, in Mr. Ford's presence, that he shouldn't make speeches like that. Mr. Ford, still furious, said, "Aw, I don't care. Don't tell him anything. I don't want him to agree with me on anything."

I told, earlier, how Mr. Ford and Edsel had had their first serious trouble over the inheritance tax, and how Mr. Ford had expressed an intention to have a board of trustees organized to carry on his policies after his death. Well, he kept changing his mind about that, but during the last phases of Edsel's illness Mr. Ford had Capizzi draw up a document setting up a board of trustees to control the company for ten years after Mr. Ford's death. The membership of the board was as follows: I was to be secretary of the board; other members were Capizzi, Sorensen, Liebold, Dr. Ruddiman, Lindbergh, Frank Campsall, and Roy Bryant, Mrs. Ford's brother. Shortly after the thing was drawn up, Mr. Ford erased Liebold's name and substituted Carl Hood, then head of Greenfield Village.

Mr. Ford had two ideas in doing this. His first idea was to prevent his heirs from ever putting the company on the market—that was the one, definite instruction he gave me in regard to the course of conduct to be followed by the board. His second motive was his wish to have the board carry on until his grandsons had matured.

However, I never had any serious idea of carrying this out. Henry became quite disturbed about the existence of this document once he came into the company. I assured him, through John Bugas, whom he sent to my office to talk to me, that it had been destroyed.

Bugas had been Edsel's prime source of information, and after Edsel died, Henry kept up his father's relationship with Bugas.

Bugas was a source of irritation to both Mr. Ford and myself. Finally I decided that the best way to cope with him was to hire him into the plant. I said to Mr. Ford, "Why don't we bring Bugas in here, and let him see what's going on?"

Mr. Ford didn't want to, but finally I persuaded him that it was the right thing to do. I thought that if Bugas were around the plant, young Henry would in time see him as I did.

Unfortunately, that didn't happen.

We brought Bugas on my staff. I gave him the title of "Administrative Assistant," because he asked for it.

He had been working for me only a few days when he wanted to fire someone. "If I could fire three or four of these fellows," he said, "the rest of them would have more respect for me." He seemed to sit around most of the day, his jacket off and his gun jutting from the shoulder holster beneath his arm, while he wrote Henry incredibly long "reports," based on material we had given him.

After a while, things got so that I expected Mr. Ford to clean house—Bugas, Henry, and all. And I firmly believe that if Mr. Ford had kept his health only a little while longer, he would have done just that.

Meanwhile, I was trying my best to like

Henry and get along with him—and not succeeding too well.

Henry seemed in haste to fire many people, even though they had served his grandfather well and long.

There was the case of Ray Rausch, for instance. When I pressed Henry for a reason for his dislike, the best he could say was: "Rausch and Harry Mack are a couple of ———."

"Why?" I said. "Because they do what your grandfather tells them to?"

Both men had been given many disagreeable tasks to perform, but they knew which side their bread was buttered on, and acted accordingly. But that had nothing to do with their value as executives. Sorensen had picked Ray Rausch and Harry Mack out of the plant and made them into executives. These two men could build a Ford car from the wheels up, and handled jobs no white-collar officials could have done.

Rausch had got very close to Sorensen during a period when Bricker was ill as the result of an automobile accident. He was capable, energetic, and inventive.

I remember, particularly, how Mr. Ford once asked Rausch to make a plastic icebox out of weeds and grass. Rausch went right to work.

I asked Mr. Ford, "Do you really mean that?"

"No," he said, "but we'll get Hub McCarroll [chief chemist] worried."

"Well," I said, "if you don't mean it you'd better stop him, because he'll sure make one."

And Rausch almost did, too—until Mr. Ford got him busy on another job.

Well, all of this hardly made me sympathetic to Henry's ideas on how things should be run around there. We had widely different conceptions of how a man ought to deal with his obligations.

If I had been sick of the atmosphere that enveloped the company when Sorensen left, my feelings now were even stronger. Mr. Ford was slipping in a serious way. Trying to cope with this was almost more of a strain than I could carry. Mr. Ford was seen at public meetings that last year, but usually he didn't know where he was—and I got that straight from his physician. There were times when he'd tell me he was going someplace when he'd just come from there.

Several times Mr. Ford took out of his pocket a bottle of pills given to him by his physicians and said, "You just find out what those doctors are giving me, Harry." And each time, to humor him, I'd check the pills at the hospital, and each time I'd tell him that the pills were phenobarbital, a sedative.

The persecution complex that he had

had all the years I knew him seemed to be growing worse, and the nightmare quality of his existence seeped into my own life. There wasn't anything I wanted but to wake up and return to a normal and healthy world.

If I had, by then, even a vestige of a desire to remain on, there occurred an incident with Bugas that would have cured me. He came to me one day to say that Henry didn't like me but that he, Bugas, had "fixed everything up."

It was appalling to think of Bugas discussing me, after the thirty years I had spent with Mr. Ford. That small incident snapped the last thread that bound me to the Ford Motor Company.

Chapter 25

THE END came suddenly. In the fall of 1945 the family arranged Mr. Ford's retirement.

Now eighty-three, and in a constantly confused state, he was kept from going out in public, was carefully guarded, and was permitted to see no one outside the family.

At the family's direction, Frank Campsall wrote out a "resignation" for Mr. Ford. In this, it was requested that Henry be made president of the company.

Mr. Ford's kitty—that safe in the laboratory whose contents I had always understood were to be mine when Mr. Ford went—was cleaned out. I was told later, by Frank Campsall, that there was a blue envelope in there, addressed to me in Mr. Ford's handwriting.

I wasn't in much of a position to do anything about the kitty or the envelope, either. There was one significant fact: This was the end of my thirty years with Henry Ford.

Shortly a Board of Directors' meeting was called. When we were assembled, they began solemnly to read Mr. Ford's "resignation." Knowing what was coming, and unable to sit silently through a farce, I got up from my chair after the first few sentences had been read and congratulated Henry.

I wanted to walk out right then, but the others prevailed on me to stay. The reading of the letter was completed, and then the directors proceeded, still with straight faces, to elect Henry president of the Ford Motor Company.

After the meeting the other directors went home. I told Henry I wanted to talk with him, and we went into his office.

As anyone else might have, I felt bitter. I told Henry, "You're taking over a billion-dollar organization here that you haven't contributed a thing to."

Then I tempered this a little, adding, "I've tried awfully hard to like you, and had hoped to part friends."

Henry said the same thing, and added, "I don't know what I'd have done without you. You know you don't have to leave—you can stay here the rest of your life."

But my mind was made up. I had always said that when Mr. Ford left I would go, and I meant it now as I had meant it before.

I walked out of the Ford Motor Company. I can't describe my feelings at leaving. The best I can say is that it was all I could do to keep myself from running down the hall, I wanted to get out so fast. I felt like a man being let out of prison.

Though I didn't officially resign for some weeks, this was my real departure date.

The gossip writers all made their conjectures on why I left—and all made wrong ones. No one there "got" me out. As evidence of this, I remained on the Ford Motor Company payroll for a year and a half after leaving. A published story to the effect that I left after a violent quarrel with Edsel's wife at a directors' meeting is an absurd lie. I met Eleanor Ford only once in my life—at an "Old-Timers" banquet at the Dearborn Inn, at which time she was very gracious to me. To the best of my knowledge, she never even attended a directors' meeting; at least, she never attended one while I was present.

I left the Ford Motor Company for just one reason: I wanted to leave.

And I wish, now, that I could make this

the last sentence of my story, and end the book here.

But, unhappily, this wasn't the end.

Although I was now separated from the company, and Mr. Ford himself was under the closest surveillance and guard, some person or persons set a gang of junior G-men to checking on me and tailing me everywhere I went.

I soon learned about this. I further learned that though I went unarmed, all the men following me around carried guns. I can't say I was much frightened.

I tired of this game, and went up to my northern Michigan ranch for a rest. There I got a message from the state police.

I was asked to call a certain *Detroit Free Press* reporter. I was to make an appointment to meet this man in Saginaw, at which time he would have in his company a man who was head of the Detroit Narcotics Squad and Harry Wismer, who was Mr. Ford's nephew and a well-known sports broadcaster.

I called the reporter and made arrangements to meet the three men. When we got together in Saginaw, the men told me why they had sought this clandestine meeting. Mr. Ford, they said, would like to see me.

I told the men to tell Mr. Ford that I had not done anything against him or said anything against him, and that I would be glad to see him.

Some time later Mr. Ford reached me by telephone. Somehow he had got to a phone outside his home.

Mr. Ford told me that he wanted me to go into the plant and shut it down. Then he began weeping and became incoherent.

I checked my impression of Mr. Ford's health with a physician who had attended him, and as a result of his opinion I paid no attention to what Mr. Ford had told me to do.

After that, Mr. Ford called me again many times. But I refused to talk to him on the phone.

On April 7, 1947, the Rouge River flooded and cut off all electric power and telephones at the Residence. In a cold room lit by oil lamps and candles, Mr. Ford died of a cerebral hemorrhage.

The next day his body lay in state in Greenfield Village, and a hundred thousand people filed past his bier.

I did not go to Mr. Ford's funeral. I was in California when I got the news. I knew that the family felt bitterly toward me, and would not welcome my presence. I knew too that if I went, the newspapers would make a show of it. So I announced that I couldn't get reservations East.

To me, Mr. Ford had died two years earlier.

Henry Ford was laid to rest in Greenfield Village. It was estimated that over

100,000 people walked through to pay him tribute. It was a tribute of simple people to a man they believed to be a hero. His formal obituaries praised him for having been an inventor and a producer. He had revolutionized American and, eventually, world industry. Perhaps his memory is most honored by the worldwide use that is made of his methods of production.